Water-in-Plants Bibliography

volume 6 1980

References no. 6606–8128 / AAS-ZOB

Editors J. Pospíšilová and J. Solárová

Dr W. Junk Publishers The Hague/Boston/London 1982

Contributors

J. Solárová
J. Pospíšilová
Z. Šesták
J. Čatsky
I. Tichá
D. Hodáňová
J. Zima

ISBN-13:978-90-6193-906-1   e-ISBN-13:978-94-009-8033-4
DOI: 10.1007/978-94-009-8033-4

PREFACE

The sixth volume of Water-in-Plants Bibliography includes papers in all fields of plant water relations research which appeared during the year 1980 - from theoretical considerations about the state of water in cells and its membrane transport to drought resistance of plants or physiological significance of irrigation. In addition to papers devoted entirely to plant water relations, papers on other topics are included if they contain data on plant hydration level, water vapour efflux, rate of water uptake or water transport, etc., or if they contain valuable methodological information (measurement of selected microclimatic factors, soil moisture, etc.).

We have tried to cover fully the relevant papers which have been published in the most important scientific periodicals and books. Articles appeared in local journals, mimeographed booklets, abstracts of thesis and of symposia contributions, etc., were chosen mostly from reprints received directly from authors. The courtesy of those authors who have already supplied us with reprints and lists of their publications is highly appreciated. The manuscript is usually prepared in May and June of the year following the year which it covers. Unfortunately some reprints come later and thus the respective references appear in the following volume, with one year delay.

To maximize the value of the bibliography the references are arranged alphabetically according to the authors' names, and each volume is provided with three indexes. The authors' index contains all names of authors, co-authors and editors. Plant genera used as experimental material are indexed according to their Latin names. The subject index covers primary items chosen according to the interest of water relations researchers. Its preparation was based not only on the titles, key words and abstracts but also on the whole content of the article.

Since more than 1500 relevant papers dealing with plant water relations and relative topics are published every year and included in this bibliography, and since all citations have been checked with the originals, collecting and preparing for publication such a large amount of material would have been impossible without the collaboration of our colleagues from the Department of Physiology of Photosynthesis and Water Relations of the Institute of Experimental Botany of the Czechoslovak Academy of Sciences in Prague. We have also acknowledge with thanks the cooperation of Mrs. Ludmila Hávová, Mrs. Lenka Kolčabová, Mrs. Marie Mandlová, Mrs. Marta Šmídová and Mrs. Drahomíra Těžká who helped in typing card material and Mrs. Zora Zawoyská and Mr. Petr Zázvorka who supplied us with rare periodicals.

Dr. Jana Pospíšilová and Dr. Jarmila Solárová

Institute of Experimental Botany
Czechoslovak Academy of Sciences

Flemingovo nám. 2
160 00 PRAHA 6
Czechoslovakia

Praha, 1 October 1981

INSTRUCTIONS FOR USE

All references are arranged alphabetically according to the authors' names. They are numbered and these numbers are used in the indexes. An asterisk preceding the number denotes the reference published in the preceding period (1975 - 1978).

Authors' names are presented in the spelling used in the original paper. If this spelling does not correspond to the spelling usually used by the author (e.g. Russian papers of English authors), one spelling is referred to the other in the Authors' Index. Like the transcriptions they are alphabetically arranged mostly according to the authors' own references. Nevertheless, the editors apologize for some misinterpretations which are partly corrected by the cross-indexing in the Authors' Index.

The references contain the original unshortened title of the paper (book). English, French, and German titles are cited in the original language. Titles in other languages are supplemented with a translation in English (using the title of the respective English abstract, if it is presented). Titles of Japanese, Chinese etc. papers are given in English translation only. In both these cases the abbreviations of the original language and the language of the abstract are given in brackets at the end of the reference. The following abbreviations are used most frequently:

Belorussian	Japanese
Bulgarian	Latvian
Chinese	Lithuanian
Croatian	Norwegian
Danish	Polish
English	Russian
Esthonian	Roumanian
French	Slovak
German	Spanish
Georgian	Swedish
Hungarian	Ukrainian
Italian	Uzbeg

The transliteration of Cyrillic characters is in accordance with the BSI-ASA/SC-Z39 draft table, i.e.:

a	а		p	п
b	б		r	р
ch	у		s	с
d	д		sh	ш
e	е		shch	щ
ё	э		t	т
f	ф		ts	ц
g	г		u	у
i	и		v	в
ĭ	й		y	ы
k	к		ya	я
kh	х		yu	ю
l	л		z	з
m	м		zh	ж
n	н		"	ъ
o	о		'	ь

Several exceptions apply for Ukrainian and Belorussian:

Ukrainian:	y	и
	i	і
	ĭ	ї
Belorussian:	ŭ	ў

The journals' names are abbreviated mainly according to the Style Manual for Biological Journals (Second Edition, Amer. Institute of Biological Sciences, Washington, D.C. 1964), e.g.:

Abhandlungen
Abstract
Abteilung
Academy
Acker
Acta
Advances
Africa (-ican)
agricultural
Agriculture
Agrobiology (-ogiya)
Agrobotanica
Agrokémia
Agronomy
agropecuaria
Akademie (-emiya)
Algology
allgemeine
Amélioration
America
American
Anais (-alele)
Analysis
analytical
Anatomy
angewandte
animal
Annales (-als)
annual
anorganic (-anisch)
applied
aquatic
Arbeit
Archiv
Argentina
Association
Atmosphere
atmospheric
atomic
Australia (-ralian)
Azerbaĭdzhanskaya
Bacteriology
Beiheft
Beiträge
Belgique
Belorusskaya
Berichte
biochemical
Biochemie
Biochemistry
biochimica
biokhimicheskiĭ
Biokhimiya
Bioklimatologie
Biologia (-ogy)
biological (-ogisk)
biophysical
Biophysics
Bodenkunde
Boletin (-ettino)
Bolgarskiĭ
botanica (-anicorum)
botanical (-anisca)
Botanika (-any)

Brasileira
Brazil
Breeding
British
Bulletin (-etins)
Byulleten
California
Canada (-adian)
cellular (-ulaire)
Center
central
Centralblatt
Československý
chemical
Chemistry
chimicus
Chinese
Chromatography
Chronicle
Ciencia
cientificas
College
Commision
Communication
comparative
Comptes Rendus
Conference
Congress
Conservation
Contamination
Contribution
Control
Croatica
cultural
Culture
current
Cytobiology
Cytochemistry
Cytology
Czechoslovak
Danske
dendrological
Dendrology
Department
Deutsche (-schland)
Development
Disease
Dissertation
Division
Doklady
Dopovidi
Drainage
ecological
Ecology
Economy
Edafology
Education
Ékologiya
eksperimental'nyĭ
Embryology
Encyclopedia
Engineering
Enology
Entomology

environmental
Enzymology
Estonskaya
European
Experiment
experimental
Faculty
Federation
Fizika
Fiziologiya
Flurbereinigung
forestiere
Forestry
Forschung
Foundation
France
Gazette
general
genetical
geneticheskiĭ
Genetics (-ika)
Geobotany
Geofizika
Geophysics
Gesellschaft
Giornale
gosudarstvennyĭ
Government
Grassland
Gruzinskaya
Helveticus
Histochemistry
Histoire (-ory)
Histology
horticultural
Horticulture
Hungaricae
Hungaricus
Husbandry
Hydrobiology
Hydrology
Indian
Industry
inorganic
Institute
Institutului
international
Investigation
Irrigation
Isotopes
issledovatel'skiĭ
Italian (-y)
Izvestiya
Jahrbuch
Japan (-anese)
Journal
Khimiya
Klasse
Kongelige
Közlemenyek
kul'turnykh
Laboratory
Landbauforschung
Landwirtschaft

lesní (-ího)
Letters
Limnology
Linnean
Litovskoĭ
Lucrarile
Magazin
Management
marina (-ine)
Material
Mathematics
Mededelingen
mediterranean
Meldinger
Meteorology
Microbiology
Midland
Mitteilungen
Modeling
modern
molecular
Monographiae (-aphy)
Moskovskiĭ (-ovskogo)
Mycology
national
natural
Naturalist
naturelle
naturkundliche
Naturforschung
nauchnye (-nyĭ)
Neerlandica
Netherland
New Zealand
Norges
Norwegian
Notiser
nuclear
Nutrition
obshcheĭ (-iĭ)
Oceanography
Oecologia
Ökologie
Optics
opytnaya (-yĭ)
organic
original
ornamental
Otdelenie
Paleobotany
Palynology
Pathology
pedagogicheskiĭ
Pesquisa
Pesticide
Pflanzen-
Pflanzenernährung
Pflanzenphysiologie
Pflanzenzüchtung
Philosophy
Photogrammetric
Phycology
physical
Physics

physiological	rolniczych	SSSR	Ukrains'kaya
Physiology	Rostlin (-lina)	Stantsii (-ntsiya)	Universidad (-ersity)
Phytologist	rostlinná	Station	US, USA
Phytopathology	Roumaine	stiintifice	USSR
Phytotaxonomy	royal	subtropicale	Uzbekskiĭ (-ekskaya)
Plantarum	Russian	summary	vědecké (-ecký)
Polonica (-ska)	Russkiĭ	Supplement	vegetable
Pollution	Sbornik	Survey	végétale
Práce	Scandinavica	Swedish	Verhandlungen
Practice	Scandinavicus	Symposium	Veröffentlichungen
prikladnoĭ	School	System	Vestnik
Proceedings	Science	Tagungsberichte	Videnskabernes
Progress	scientific	technical (-nische)	Virology
Publication	Section	Technology	Virusforschungen
Publlishers	Selektsiya	Tekhnika	Viticulture
Quality	Selskabs	theoretical	Volume
quantitative	Sel'skokhozyaĭstvo	thermal	Voprosy
Quarterly	Series (-iya)	Tidsskrift	vostochnyĭ
Radiation	Service	Tijdschrift	vsesoyuznyĭ
Radiobiology	Shkoly (-oly)	Toxicology	vyssheĭ (-iĭ)
Rasteniĭ	Sibirskiĭ (-skogo)	Transactions	výzkumný (-umného)
Rastenievodstvo	Skrifter	Travail (-aux)	Weekblatt
Recherche (-erches)	Slovak (-enská)	tropical (-icale)	Wetenschappen
Report	Society	Trudy	Wissenschaft
Research	Soobshcheniya	Turkmenskaya	Zapiski
Resources	Sovetskiĭ (-iet)	uchenye	Zeitschrift
Review (-ista, -ue)	sovremennyĭ	Ugeskrift	Zeitung
Rivista	special	United Kingdom	Zentralblatt
Roczniky	sperimentale	Ukraïnian	Zhurnal

The numbers at the end of each reference of a journal article denote: volume (issue) : first page - last page, year of publication. The number of issue is given only in journals where each issue is paginated separately.

Book titles are cited according to the title page, not to the book jacket or cover. The publishing house, place and year of publication are included.

Printers' errors in the original papers are marked by underlining the respective words (letters).

6606 - AASE, J.K., SIDDOWAY, F.H.: Stubble height effects on seasonal microclimate, water balance, and plant development of no-till winter wheat. - Agr. Meteorol. 21: 1-20, 1980.

6607 - ABROL, B.K., MACKAY, D.B.: Plumular abnormality in wheat seed. - Seed Sci. Technol. 8: 59-76, 1980.

6608 - ABRUÑA, F., VICENTE-CHANDLER, J., IRIZARRY, H., SILVA, S.: Evapotranspiration with plantains and the effect of frequency of irrigation on yields. - J. agr. Univ. Puerto Rico 64: 204-210, 1980.

*6609 - ACEVEDO, E., FERERES, E., HSIAO, T.C., HENDERSON, D.W.: Diurnal growth trends, water potential, and osmotic adjustment of maize and sorghum leaves in the field. - Plant Physiol. 64: 476-480, 1979.

*6610 - ACHARYA, C.L., SANDHU, S.S., ABROL, I.P.: Effect of exchangeable sodium on the rate and pattern of water uptake by raya (*Brassica juncea* L.) in the field. - Agron. J. 71: 936-941, 1979.

6611 - ACKERSON, R.C.: Stomatal response of cotton to water stress and abscisic acid as affected by water stress history. - Plant Physiol. 65: 455-459, 1980.

6612 - ACKERSON, R.C., KRIEG, D.R., SUNG, F.J.M.: Leaf conductance and osmoregulation of field-grown sorghum genotypes. - Crop. Sci. 20: 10-14, 1980.

6613 - ACKLEY, W.B., KRUEGER, W.H.: Overhead irrigation water quality and the cracking of sweet cherries. - HortScience 15: 289-290, 1980.

6614 - ADAMS, D.E., ANDERSON, R.C.: Species response to a moisture gradient in central Illinois forests. - Amer. J. Bot. 67: 381-392, 1980.

6615 - ADDICOT, F.T.: Introductory comments: Abscisic acid in the physiology of plants. - In: SKOOG, F. (ed.): Plant Growth Substances 1979. Pp. 241. Springer-Verlag, Berlin - Heidelberg - New York 1980.

*6616 - AGABBIO, M.: Influenza dell'intervento irriguo sul ciclo produttivo dell'olivo. Nota II: Influenza del regime idrico sulla biologia floreale e sui caratteri morfo-qualitativi dei frutti. [The effect of irrigation intervals and volumes on productivity of olive trees. Part II: Influence of water balance on floral biology, morphological and qualitative characteristics of fruits.] - Stud. Sassar. Sez. III. Ann. Fac. Agr. Univ. Sassari 25: 3-9, 1978. [In Ital, ab: E.]

*6617 - AGGARWAL, R.K., SINGH, P.: Effect of Zn and P levels on the concentration and uptake of N and N/Zn ratio in rainfed pearl millet (*Pennisetum typhoides* S and H). - Ann. Arid Zone 17: 267-272, 1978.

6618 - AHMED, A.M., HEIKAL, M.M., ZIDAN, M.A.: Effects of salinization treatments on growth and some related physiological activities of some leguminous plants. - Can. J. Plant Sci. 60: 713-720, 1980.

6619 - AHMED, S., FLETCHER, R.A.: Reduced transpiration and increased water efficiency by diuron in corn (*Zea mays*). - Weed Sci. 28: 180-185, 1980.

6620 - AHO, N.: Effet de la sécheresse atmosphérique sur la photosynthèse nette de deux espèces de type C_4: le Sorgho et le Maïs. - C.R. Acad. Sci. Paris Sér. D 290: 543-546, 1980.

6621 - AHO, N., DAUDET, F.-A., VARTANIAN, N.: Variations de la réserve facilement utilisable en eau du sol en relation avec différents facteurs écologiques. - Ann. agron. 31: 109-124, 1980.

6622 - AKALEHIYWOT, T., BEWLEY, J.D.: Desiccation of oat grains during and following germination, and its effects upon protein synthesis. - Can. J. Bot. 58: 2349-2355, 1980.

6623 - AKITA, S.: [Studies on the differences in photosynthesis and photorespiration among crops I. The differential responses of photosynthesis, photorespiration and dry matter production to oxygen concentration among species.] - Bull. nat. Inst. agr. Sci., Ser. D 31: 1-58, 1980. [In Jap, ab: E.]

6624 - AKITA, S.: [Studies on the differences in photosynthesis and photorespiration among crops II. The differential responses of photosynthesis, photorespiration and dry matter production to carbon dioxide concentration among species.] - Bull. nat. Inst. agr. Sci., Ser. D 31: 59-94, 1980. [In Jap, ab: E.]

*6625 - ALBERS, D.J., CARPENTER, S.B.: Influence of site, environmental conditions, mulching, and herbaceous ground cover on survival, growth, and water relations of european alder seedlings planted on surface mine spoil. - In: CARPENTER, S. B. (ed.): Symposium on Surface Mining Hydrology, Sedimentology and Reclamation. Pp. 23-32. Ores Publications, Lexington 1979.

6626 - ALBERT, R., KÖNIGSHOFER, H., KINZEL, H.: Zur Osmoregulation einer physiologisch calciophoben und ökologisch calcicolen Pflanze (*Dianthus lumnitzeri* Wiesb.). - Flora 169: 9-14, 1980.

*6627 - ALBINET, E., FURCILĂ, P., JITCĂ, G., SECĂREANU, G.: Consumul de apă calculat după metoda Thornthwaite la principalele culturi de cîmp pentru condiţiile cîmpiei Siretului inferior. [Water consumption calculated according to the Thornthwaite method for the main field crops in the conditions of the Siret Lower course plain.] - Lucrări ştiinţ. Ser. agron 1978: 11-14, 1978. [In Roum, ab: E.]

*6628 - ALBREGTS, E.E., HOWARD, C.M.: Influence of fertilizer sources and drip irrigation on strawberries. - Soil Crop Sci. Soc. Florida Proc. 37: 159-162, 1977.

*6629 - ALESSI, J., POWER, J.F., SIBBITT, L.D.: Yield, quality and nitrogen fertilizer recovery of standard and semidwarf spring wheat as affected by sowing date and fertilizer rate. - J. agr. Sci. 93: 87-93, 1979.

*6630 - ALI, H.C., WILLIAMS, R.L., JOHNSON, M.W.,Jr.: The relationships of leaf area to grain yield and other factors in corn (*Zea mays* L.). - Z. Pflanzenzücht. 80: 320-325, 1978.

6631 - AL-ITHAWI, B., DEIBERT, E.J., OLSON, R.A.: Applied N and moisture level effects on yield, depth of root activity, and nutrient uptake by soybeans. - Agron. J. 72: 827-832, 1980.

6632 - ALLEN, J.J., NELL, T.A., JOINER, J.N., ALBRIGO, L.G.: Effects of leaf position and storage conditions on pressure bomb measurement of leaf water potential in chrysanthemums. - HortScience 15: 808-809, 1980.

6633 - ALLEN, R.R., MUSICK, J.T., DUSEK, D.A.: Limited tillage and energy use with furrow-irrigated grain sorghum. - Trans. ASAE 23: 346-350, 1980.

*6634 - ALONI, B., PRESSMAN, E.: Petiole pithiness in celery leaves: Induction by environmental stresses and the involvement of abscisic acid. - Physiol. Plant. 47: 61-65, 1979.

*6635 - ALPERT, P.: Desiccation of desert mosses following a summer rainstorm. - Bryologist 82: 65-71, 1979.

6636 - ALSTON, A.M.: Response of wheat to deep placement of nitrogen and phosphorus fertilizers on a soil high in phosphorus in the surface layer. - Aust. J. agr. Res. 31: 13-24, 1980.

*6637 - ALVIM, P.de T.: Cacao. - In: ALVIM, P.de T., KOZLOWSKI, T.T. (ed.): Ecophy-
siology of Tropical Crops. Pp. 279-313. Academic Press, New York - San Francisco - London 1977.

*6638 - ALVIM, P.de T., KOZLOWSKI, T.T. (ed.): Ecophysiology of Tropical Crops. -
Academic Press, New York - San Francisco - London 1977.

*6639 - ALZUBAIDI, A.H., EL-BASSAM, N.: Use of radioactive sodium-22 to study the
processes of soil salinization and desalinization. - In: Isotopes and Radiation in Research on Soil-Plant Relationships. Pp. 259-270. International
Atomic Energy Agency, Vienna 1979.

 6640 - ANDALES, S.C., PETTIBONE, C.A., DAVIS, D.C.: Influence of relative humidity
and temperature on weight and sucrose losses of stored sugarbeets. - Trans.
ASAE 23: 477-480, 1980.

*6641 - ANDERSON, J.E.: Transpiration and photosynthesis in saltcedar. - Hydrol.
Water Resour. Arizona Southwest 7: 125-131, 1977.

*6642 - ANDERSON, W.K.: Water use of sunflower crops. - Aust. J. exp. Agr. anim. Husb.
19: 233-240, 1979.

 6643 - ANDERSON, W.K.: Some water use response of barley, lupine and rapeseed. -
Aust. J. exp. Agr. anim. Husb. 20: 202-209, 1980.

 6644 - ANGUILLESI, M.C., GRILLI, I., FLORIS, C.: Rate of synthesis of spermine and
spermidine in germinating seeds of *Glycine, Helianthus* and *Triticum*. -
Planta 148: 24-27, 1980.

 6645 - ANGUS, J.F., NIX, H.A., RUSSELL, J.S., KRUIZINGA, J.E.: Water use, growth
and yield of wheat in a subtropical environment. - Aust. J. agr. Res. 31:
873-886, 1980.

 6646 - ANTIPOV, N.I.: Vodnyĭ rezhim gallov list'ev raznykh vidov rasteniĭ. [Water
relations in leaf galls of various plant species.] - Fiziol. Rast. 27: 140-
144, 1980. [In R, ab: E.]

*6647 - ANTLFINGER, A.E., DUNN, E.L.: Seasonal patterns of CO_2 and water vapor
exchange of three salt marsh succulents. - Oecologia 43: 249-260, 1979.

 6648 - APARICIO-TEJO, P.M., SÁNCHEZ-DÍAZ, M.F., PEÑA, J.I.: Nitrogen fixation, stomatal response and transpiration in *Medicago sativa, Trifolium repens* and
T. subterraneum under water stress and recovery. - Physiol. Plant. 48: 1-4,
1980.

 6649 - APARICIO-TEJO, P.M., SÁNCHEZ-DÍAZ, M.F., PEÑA, J.I.: Measured and calculated
transpiration in *Trifolium repens* under different water potentials. - J. exp.
Bot. 31: 839-843, 1980.

*6650 - APEL, P.: Leitbündeldichte und Stomatafrequenz von Gramineen-Arten mit C_3-
beziehungsweise C_4-pathway der Photosynthese. - Kulturpflanze 27: 91-95,
1979.

*6651 - APPELBAUM, S., SCOTT, T.K.: Agricultural management strategies, initiatives,
and goals for survival. - In: SCOTT, T.K. (ed.): Plant Regulation and World
Agriculture. Pp. 511-530. Plenum Press, New York - London 1979.

*6652 - ARDITTI, J., HARRISON, C.R.: Postpollination phenomena in orchid flowers.
VIII. Water and dry weight relations. - Bot. Gaz. 140: 133-137, 1979.

*6653 - ARTYKOV, K.: Agrometeorologicheskie usloviya prizhivaemosti vskhodov kustar-
nikov v predgor'yakh yugo-vostochnogo Turkmenistana. [Agrometeorology and
taking root of shrubs in piedmont of southeastern Turkmenistan.] - Problemy
Osvoeniya Pustyn' 4: 78-82, 1975. [In R, ab: Turkm,E.]

6654 - ASADA, S., TAKANO, M., SHIBASAKI, I.: Mutation induced by drying of *Escherichia coli* on a hydrophobic filter membrane. - Appl. environ. Microbiol. 40: 274-281, 1980.

*6655 - ASHCROFT, W.J., MURRAY, D.R.: The dual functions of the cotyledons of *Acacia iteaphylla* F. Muell. (*Mimosoideae*). - Aust. J. Bot. 27: 343-352, 1979.

6656 - ASPINALL, D.: Role of abscisic acid and other hormones in adaptation to water stress. - In: TURNER, N.C., KRAMER, P.J. (ed.): Adaptation of Plants to Water and High Temperature Stress. Pp. 155-172. John Wiley & Sons, New York - Chichester - Brisbane - Toronto 1980.

*6657 - ASTON, A.R.: Rainfall interception by eight small trees. - J. Hydrol. 42: 383-396, 1979.

6658 - ATTIWILL, P.M., CLOUGH, B.F.: Carbon dioxide and water vapour exchange in the white mangrove. - Photosynthetica 14: 40-47, 1980.

*6659 - AUGSPURGER, C.K.: Irregular rain cues and the germination and seedling survival of a Panamanian shrub (*Hybanthus prunifolus*). - Oecologia 44: 53-59, 1979.

6660 - AUSSENAC, G.: Comportement hydrique de rameaux excisés de quelques espèces de sapins et de pins noirs en phase de dessiccation. - Ann. Sci. forest. 37: 201-215, 1980.

6661 - AUSSENAC, G., BOULANGEAT, C.: Interception des précipitations et évapotranspiration réelle dans des peuplements de feuillu (*Fagus silvatica* L.) et de résineux (*Pseudotsuga menziesii* (Mirb) Franco). - Ann. Sci. forest. 37: 91-107, 1980.

6662 - AYRES, P.G.: Stomatal behaviour in mildewed pea leaves: Solute potentials of the epidermis and effects of pisatin. - Physiol. Plant Pathol. 17: 157-165, 1980.

6663 - AYRES, P.G., WOOLACOTT, B.: Effects of soil water level on the development of adult plant resistance to powdery mildew in barley. - Ann. appl. Biol. 94: 255-263, 1980.

*6664 - BACON, G.J., BACHELARD, E.P.: The influence of nursery conditioning treatments on some physiological responses of recently transplanted seedlings of *Pinus caribaea* Mor. var. *hondurensis* B.& G. - Aust. Forest Res. 8: 171-183, 1978.

*6665 - BACON, G.J., BACHELARD, E.P.: Changes in growth substance levels associated with the conditioning of *Pinus caribaea* Mor. var. *hondurensis* B.& G. seedlings to water stress. - Aust. Forest Res. 9: 241-254, 1979.

*6666 - BAGNALL, D.: Low temperature responses of three *Sorghum* species. - In: LYONS, J.M., GRAHAM, D., RAISON, J.K. (ed.): Low Temperature Stress in Crop Plants. The Role of the Membrane. Pp. 67-80. Academic Press, New York 1979.

6667 - BAILISS, K.W., PLAZA-MORALES, G.: Effects of postinoculation leaf water status on infection of French bean by tobacco necrosis virus. - Physiol. Plant Pathol. 17: 357-367, 1980.

6668 - BAKER, D.N.: Simulation for research and crop management. - In: CORBIN, F.T. (ed.): World Soybean Research Conference II: Proceedings. Pp. 533-546. Westview Press, Boulder 1980.

6669 - BAKER, E.A., PROCOPIOU, J.: Effect of soil moisture status on leaf surface wax yield of some drought-resistant species.- J. hort. Sci. 55: 85-87, 1980.

6670 - BAKRADZE, N.G., BALLA, Yu.I., METREVELI, I.M., SHARIMANOV, Yu.G.: Orientatsionnaya zavisimost' liniĭ protonnogo rezonansa vody v kapillyarakh steblya rasteniĭ. [Orientation relationship of the lines of water proton resonance in plant stem capillaries.] - Biofizika 25: 356-357, 1980. [In R, ab: E.]

6671 - BAKRADZE, N.G., MOISTSRAPISHVILI, K.M., KESHELASHVILI, L.V.: Kristallizatsiya vody v rastitel'nykh ob"ektakh. [Water crystallization in plant specimens.] - Biofizika 25: 54-58, 1980. [In R, ab: E.]

6672 - BALINA, N.V.: Vliyanie atmosfernoĭ zasukhi na protsess meĭoza u fasoli. [Effect of atmospheric drought on meiosis in kidney bean plants.] - Fiziol. Rast. 27: 272-277, 1980. [In R, ab: E.]

*6673 - BANERJI, M.L., MAITY, G.G., HORE, D.K.: Giant stomata of some angiosperm taxa. - Acta Bot. Ind. 7: 185-187, 1979.

*6674 - BANGE, G.: Comparison of compartmental analysis for Rb^+ and Na^+ in low- and high-salt barley roots. - Physiol. Plant. 46: 179-183, 1979.

6675 - BAŇOCH, Z., HÁJEK, L.: Cesty k optimalizaci hnojení v závlahových podmínkách. [Optimizing fertilization in irrigated fields.] - Rost. Výroba (Praha) 26: 799-806, 1980. [In Czech, ab: E,R,G.]

6676 - BAŇOCH, Z., RASZKA, P.: Možnost racionalizace závlahového režimu ozimé pšenice použitím zásobní závlahy. [Possibilities of rationalizing irrigation regimes of winter wheat using store irrigation.] - Rost. Výroba (Praha) 26: 575-582, 1980. [In Czech, ab: R,E,G.]

6677 - BANSAL, R.P., BHATI, P.R., SEN, D.N.: Differential specificity in water imbibition of Indian arid zone seeds. - Biol. Plant. 22: 327-331, 1980.

6678 - BARASSI, C.A., CRUPKIN, M., SUELDO, R.J., INVERNATI, A.: Osmotic stress in coleoptiles and primary leaves of wheat I. Effects on size, weight, total protein, DNA, and phosphorus. - J. exp. Bot. 31: 1565-1572, 1980.

6679 - BARBER, J., NAKATANI, H.Y.: Techniques for studying ionic regulations of chloroplasts. - In: COLOWICK, S.P., KAPLAN, N.O. (ed.): Methods in Enzymology. Vol. 69. Pp. 585-604. Academic Press, New York - London - Toronto - Sydney - San Francisco 1980.

6680 - BARDEN, J.A., LOVE, J.M., PORPIGLIA, P.J., MARINI, R.P., CALDWELL, J.D.: Net photosynthesis and dark respiration of apple leaves are not affected by shoot detachment. - HortScience 15: 595-597, 1980.

6681 - BARLOW, E.W.R., LEE, J.W., MUNNS, R., SMART, M.G.: Water relations of the developing wheat grain. - Aust. J. Plant Physiol. 7: 519-525, 1980.

6682 - BARLOW, E.W.R., MUNNS, R.E., BRADY, C.J.: Drought responses of apical meristems. - In: TURNER, N.C., KRAMER, P.J. (ed.): Adaptation of Plants to Water and High Temperature Stress. Pp. 191-205. John Wiley & Sons, New York - Chichester - Brisbane - Toronto 1980.

*6683 - BARNES, J.E., REID, R.E.: Effect of mid-season draining on paddy rice in the Lower Burdekin Valley. - Queensland J. agr. anim. Sci. 35: 159-167, 1978.

*6684 - BARTHOLOMEW, D.P., KADZIMIN, S.B.: Pineapple. - In: ALVIM, P.de T., KOZLOWSKI, T.T. (ed.): Ecophysiology of Tropical Crops. Pp. 113-156. Academic Press, New York - San Francisco - London 1977.

6685 - BARTON, J.R., McLAUGHLIN, S.B., McCONATHY, R.K.: The effects of SO_2 on components of leaf resistance to gas exchange. - Environ. Pollut. Ser. A: Ecol. Biol. 21: 255-265, 1980.

6686 - BARUAH, K.K., SINGH, O.S.: Effect of moisture-regimes & iron-levels on chlorosis & ion-uptake by paddy *Oryza sativa* L. - Ind. J. exp. Biol. 18: 184-187, 1980.

6687 - BAR-YOSEF, B., STAMMERS, C., SAGIV, B.: Growth of trickle-irrigated tomato as related to rooting volume and uptake of N and water. - Agron. J. 72: 815-822, 1980.

*6688 - BASU, R.N.: Physico-chemical control of seed deterioration. - Seed Res. 4: 15-23, 1976.

*6689 - BASU, R.N., PAL, P.: Physicochemical control of seed deterioration in rice. - Ind. J. agr. Sci. 49: 1-6, 1979.

6690 - BAUDER, J.W., MONTGOMERY, B.R.: N-source and irrigation effects on nitrate leaching. - Agron. J. 72: 593-596, 1980.

6691 - BAUER, H., BAUER, U.: Photosynthesis in leaves of the juvenile and adult phase of ivy (*Hedera helix*). - Physiol. Plant. 49: 366-372, 1980.

*6692 - BEARDSELL, D.V., NICHOLS, D.G., JONES, D.L.: Water relations of nursery potting-media. - Sci. Hort. 11: 9-17, 1979.

*6693 - BEDUNAH, D., TRLICA, M.J.: Sodium chloride effects on carbon dioxide exchange rates and other plant and soil variables of ponderosa pine. - Can. J. Forest Res. 9: 349-353, 1979.

6694 - BEER, S., ESHEL, A., WAISEL, Y.: Carbon metabolism in seagrasses. III. Activities of carbon fixing enzymes in relation to internal salt concentrations. - J. exp. Bot. 31: 1027-1033, 1980.

6695 - BEESE, F., WIERENGA, P.J.: Solute transport through soil with adsorption and root water uptake computed with a transient and a constant-flux model. - Soil Sci. 129: 245-252, 1980.

6696 - BEGG, J.E.: Morphological adaptations of leaves to water stress. - In: TURNER, N.C., KRAMER, P.J. (ed.): Adaptation of Plants to Water and High Temperature Stress. Pp. 33-42. John Wiley & Sons, New York - Chichester - Brisbane - Toronto 1980.

*6697 - BÉGUIN, C.: Pressions de sève le long d'un transect marécageux; Essai d'interprétation phytosociologique et synécologique. - Bull. Soc. Neuchateloise Sci. Natur. 102: 81-103, 1979.

*6698 - BEHBOUDIAN, M.H.: Responses of eggplant to drought. II. Gas exchange parameters. - Sci. Hort. 7: 311-317, 1977.

*6699 - BEINEKE, W.E., HUNLEY, J.L.: Delay of floral and leaf development by over-tree irrigation in black walnut. - Can.J. Forest Res. 9: 379-382, 1979.

*6700 - BELAN, F.: Změny obsahu vody v půdě ve vegetačním období. [Changes in soil water content during seasonal course.] - In: Bilancia Energie a Vody v Polných a Lesných Ekosystémoch. Pp. 70-78. Vysoká Škola Polnohospodárská, Nitra 1979. [In Czech.]

*6701 - BELAYA, G.A.: Vodnyǐ rezhim krupnotrav'ya Kamchatki v razlichnykh ėkologicheskikh usloviyakh. [Water relations of large meadows in Kamchatka in different ecological conditions.] - Ėkologiya 1978 (2): 73-75, 1978. [In R.]

*6702 - BELAYA, G.A., MOROZOV, V.L.: Ėkologo-fiziologicheskaya kharakteristika subal'piiskogo krupnotrav'ya na Kamchatke. [Ecophysiological characteristics of subalpine broadleaf ecosystems in Kamchatka.] - Ekol. Biol. Vysokogornykh Rast. 14 (2): 17-23, 1979. [In R.]

6703 - BELFORD, R.K., CANNELL, R.Q., THOMSON, R.J., DENNIS, C.W.: Effects of waterlogging at different stages of development on the growth and yield of peas
(*Pisum sativum* L.). - J. Sci. Food Agr. 31: 857-869, 1980.

*6704 - BELL, K.L., BLISS, L.C.: Root growth in a polar semidesert environment. -
Can. J. Bot. 56: 2470-2490, 1977.

*6705 - BELL, K.L., BLISS, L.C.: Autecology of *Kobresia bellardii*: Why winter snow
accumulation limits local distribution. - Ecol. Monogr. 49: 377-402, 1979.

*6706 - BELOT, Y., GAUTHIER, D., CAMUS, H., CAPUT, C.: Prediction of the flux of
tritiated water from air to plant leaves. - Health Physics 37: 575-583, 1979.

*6707 - BELT, G.H., KING, J.G.: Augumenting summer streamflow by use of a silicone
antitranspirant. - Water Resour. Res. 13: 267-272, 1977.

6708 - BEN-AMOTZ, A.: Glycerol production in the alga *Dunaliella*. - In: SAN PIETRO,
A. (ed.): Biochemical and Photosynthetic Aspects of Energy Production. Pp.
191-208. Academic Press, New York - London - Toronto - Sydney - San Francisco
1980.

*6709 - BEN-AMOTZ, A., AVRON, M.: On the mechanism of osmoregulation in *Dunaliella*. -
In: CAPLAN, S.R., GINZBURG, M. (ed.): Energetics and Structure of Halophilic
Microorganisms. Pp. 529-541. Elsevier - North-Holland Biomedical Press,
Amsterdam - New York 1978.

6710 - BEN-AMOTZ, A., AVRON, M.: Osmoregulation in the halophilic algae *Dunaliella*
and *Asteromonas*. - In: RAINS, D.W., VALENTINE, R.C., HOLLAENDER, A. (ed.):
Genetic Engineering of Osmoregulation. Impact on Plant Productivity for Food,
Chemicals, and Energy. Pp. 91-99. Plenum Press, New York - London 1980.

*6711 - BENECKE, P., Van der PLOEG, R.R.: Quantifizierung des zeitlichen Verhaltens
der Wasserhaushaltskomponenten eines Buchen- und eines Fichtenaltholzbestan-
des im Solling mit hilfe bodenhydrologischer Methoden. - Verh. Ges. Ökol.
1976: 3-15, 1976.

6712 - BENECKE, U.: Photosynthesis and transpiration of *Pinus radiata* D. Don under
natural conditions in a forest stand. - Oecologia 44: 192-198, 1980.

6713 - BENOIT, G.R., GRANT, W.J.: Plant water deficit effects on Aroostook County
potato yields over 30 years. - Amer. Potato J. 57: 585-594, 1980.

6714 - BENTLEY, B.L., CARPENTER, E.J.: Effects of desiccation and rehydration on
nitrogen fixation by epiphylls in a tropical rainforest. - Microbiol. Ecol.
6: 109-113, 1980.

6715 - BERKALOFF, C., DUVAL, J.C.: Correlated enfluence of cation concentration and
excitation intensity on PS II activity - II. Comparative study between green
plant and brown-alga chloroplasts. - Photosynthesis Res. 1: 127-135, 1980.

6716 - BERKOWITZ, G.A., HOPPER, N.W.: A method of increasing the accuracy of diffusive resistance porometer calibrations. - Ann. Bot. 45: 723-727, 1980.

6717 - BERRY, J., BJÖRKMAN, O.: Photosynthetic response and adaptation to temperature in higher plants. - Annu. Rev. Plant Physiol. 31: 491-543, 1980.

6718 - BESRI, M.: Influence du potentiel osmotique de l'eau sur la croissance de
Fusarium oxysporum f.sp. *lycopersici* et de *Verticillium dahliae*. - Phytopathol. Z. 99: 1-8, 1980.

6719 - BETTS, M.F., MORRISON, I.N.: Effects of chemical desiccation versus swathing on seed yield and quality of fababeans (*Vicia faba*). - Can. J. Plant Sci. 60: 1115-1121, 1980.

*6720 - BEVERSDORF, W.D.: Influence of ploidy level on several plant characteristics in soybeans. - Can. J. Plant Sci. 59: 945-948, 1979.

6721 - BEWLEY, J.D., LARSEN, K.M.: Cessation of protein synthesis in water-stressed pea roots and maize mesocotyls without loss of polyribosomes. Effect of letha and non-lethal water stress. - J. exp. Bot. 31: 1245-1256, 1980.

*6722 - BHARATI, M.P.: Density and environment interaction in soybean: a review. - Nepalese J. Agr. 12: 239-247, 1977.

6723 - BIBLE, B.B., JU, H.-Y., CHONG, C.: Influence of cultivar, season, irrigation, and date of planting on thiocyanate ion content in cabbages. - J. Amer. Soc. hort. Sci. 105: 88-91, 1980.

6724 - BIDDINGTON, N.L., THOMAS, T.H., DEARMAN, A.S.: The promotive effect on subsequent germination of treating imbibed celery seeds with high temperature before or during drying. - Plant Cell Environ. 3: 461-465, 1980.

6725 - BIDINGER, F.: Breeding for drought resistance. - In: TURNER, N.C., KRAMER, P.J. (ed.): Adaptation of Plants to Water and High Temperature Stress. Pp. 452-454. John Wiley & Sons, New York - Chichester - Brisbane - Toronto 1980.

6726 - BIKHELE, Z.N., MOLDAU, Kh.A., ROSS, Yu.K.: Matematicheskoe Modelirovanie Transpiratsii i Fotosinteza Rastenii pri Nedostatke Pochvenoi Vlagi. [Mathematical Modelling of Transpiration and Photosynthesis in Plants Affected by by Soil Moisture Stress.] - Gidrometeoizdat, Leningrad 1980. [In R, ab: E.]

6727 - BILLARD, J.P., BOUCAUD, J.: Effect of NaCl on the activities of glutamate synthase from a halophyte *Suaeda maritima* and from a glycophyte *Phaseolus vulgaris*. - Phytochemistry 19: 1939-1942, 1980.

*6728 - BINDEROVÁ, A.: Vliv teploty a srážek na výnosy sušiny vojtěšky seté. [Effect of temperature and rainfall on lucerne yields.] - Rost. Výroba (Praha) 25: 313-322, 1979. [In Czech, ab: E,G,R.]

6729 - BINGHAM, G.E., COYNE, P.I., KENNEDY, R.B., JACKSON, W.L.: Design and Fabrication of a Portable Minicuvette System for Measuring Leaf Photosynthesis and Stomatal Conductance under Controlled Conditions. - Lawrence Livermore Laboratory, Livermore 1980.

*6730 - BIRYUKOV, V.N., MALAN'IN, A.N., BIRYUKOVA, Z.P., SOLOV' EV, A.M.: Prichiny usykhaniya sosny v lentochnykh borakh priirtysh'ya. [Causes of drying-up of pine in ribbon pine forests of the Irtysh river region.] - Lesovedenie 1979 (2): 3-12, 1979. [In R, ab: E.]

6731 - BJÖRKMAN, O., BADGER, M.R., ARMOND, P.A.: Response and adaptation of photosynthesis to high temperatures. - In: TURNER, N.C., KRAMER, P.J. (ed.): Adaptation of Plants to Water and High Temperature Stress. Pp. 233-249. John Wiley & Sons, New York - Chichester - Brisbane - Toronto 1980.

6732 - BJÖRKMAN, O., DOWNTON, W.J.S., MOONEY, H.A.: Response and adaptation to water stress in *Nerium oleander*. - Carnegie Inst. Washington Year Book 79: 150-157, 1980.

6733 - BLACK, V.J., UNSWORTH, M.H.: Stomatal responses to sulphur dioxide and vapour pressure deficit. - J. exp. Bot. 31: 667-677, 1980.

6734 - BLACKLOW, W.M., MAYBURY, K.G.: A battery-operated instrument for non-destruc-
tive measurements of photosynthesis and transpiration of ears and leaves of
cereals using $^{14}CO_2$ and a lithium chloride hygrometer. - J. exp. Bot. 31:
1119-1129, 1980.

6735 - BLACKWELL, P.S., ELSWORTH, M.J.: A system for automatically measuring and
recording soil water potential and rainfall. - Agr. Water Manage. 3: 135-141,
1980.

*6736 - BLAKE, J., ZAERR, J., HEE, S.: Controlled moisture stress to improve cold
hardiness and morphology of douglas-fir seedlings. - Forest Sci. 25: 576-582,
1979.

6737 - BLAKE, T.J.: Effect of coppicing on growth rates, stomatal characteristics
and water relations in *Eucalyptus camaldulensis* Dehn. - Aust. J. Plant Physiol.
7: 81-87, 1980.

*6738 - BLANCHET, R., GELFI, N.: Relations entre développement foliaire, transpiration
et production chez le soja (cv Amsoy 71 et Hodgson). - Ann. agron. 29: 223-
242, 1978.

6739 - BLANCHET, R., GELFI, N.: Caractères xérophytiques de quelques espèces d'*Helianthus* susceptibles d'être utilisés pour améliorer l'adaptation aux conditions sèches du Tournesol cultivé (*Helianthus annuus* L.). - C.R. Acad. Sci.
Paris, Sér. D. 290: 279-282, 1980.

6740 - BLECKMANN, C.A., HULL, H.M., HOSHAW, R.W.: Cuticular ultrastructure of
Prosopis velutina and *Acacia greggi* leaflets. - Bot. Gaz. 141: 1-8, 1980.

*6741 - BLEKHMAN, G.I.: Kolichestvennye izmeneniya v soderzhanii ribonukleozidmono-
fosfatov v tsitoplazme list'ev pshenitsy pri obezvozhivanii. [Quantitative
changes in the content of ribonucleoside monophosphates in wheat leaf cyto-
plasm as affected by plant dehydration.] - Fiziol. Rast. 26: 779-787, 1979.
[In R, ab: E.]

6742 - BLESSINGTON, T.M., SRITHAVAJ, A.A., SINGLETARY, C.C.: Influence of watering
methods and ancymidol on two tropical foliage species held in a simulated
interior environment. - J. Amer. Soc. hort. Sci. 105: 785-787, 1980.

6743 - BLUM, A.: Genetic improvement of drought adaptation. - In: TURNER, N.C.,
KRAMER, P.J. (ed.): Adaptation of Plants to Water and High Temperature
Stress. Pp. 450-452. John Wiley & Sons, New York - Chichester - Brisbane -
Toronto 1980.

*6744 - BLUNDEN, G., JONES, E.M., PASSAM, H.C., METCALF, E.: Increases in chlorophyll
retention times of limes after post-harvest immersion in N_6-benzyladenine and
gibberellic acid. - Trop. Agr. 56: 311-319, 1979.

6745 - BLUNT, C.G., JONES, R.J.: The use of leaf development rate to determine time
to irrigate pangola grass. - Aust. J. exp. Agr. anim. Husb. 20: 556-560, 1980.

6746 - BOGGESS, S.F., STEWART, C.R.: The relationship between water stress induced
proline accumulation and inhibition of protein synthesis in tobacco leaves. -
Plant Sci. Lett. 17: 245-252, 1980.

*6747 - BOHRA, D.R., SONI, S.R., SHARMA, B.D.: Ferns of Rajasthan - behaviour of
chlorophyll and carotenoids in drought resistance. - Experientia 35: 332-333,
1979.

*6748 - BOÏKOV, S.: Vodorazkhod na ljutsernata za furazh v Kraïdunavskata nizina na
Severozapadna B''lgariya. [Alfalfa evapotranspiration in the Danube plain of
Northwestern Bulgaria.] - Rasteniev. Nauki 16 (8): 59-69, 1979. [In Bulg, ab:
E,R.]

6749 - BOLE, J.B., PITTMAN, U.J.: Spring soil water, precipitation, and nitrogen
       fertilizer: Effect on barley yield. - Can. J. Soil Sci. 60: 461-469, 1980.

6750 - BOLE, J.B., PITTMAN, U.J.: Spring soil water, precipitation, and nitrogen
       fertilizer: Effect on barley grain protein content and nitrogen yield. -
       Can. J. Soil Sci. 60: 471-477, 1980.

*6751 - BOLE, J.B., WELLS, S.A.: Dryland soil salinity: Effect on the yield and yield
       components of 6-row barley, 2-row barley, wheat, and oats. -'Can. J. Soil Sci.
       59: 11-17, 1979.

*6752 - BOLOGA, A.S.: Experiments on the photosynthesis rate in the alga *Cladophora
       vagabunda* L. under modified salinity and ionic ratios. - Rev. Roum. Biol. -
       Biol. vég. 24: 127-131, 1979.

*6753 - BONNER, F.T.: Measurement of seed moisture in *Liriodendron, Prunus,* and *Pinus.*
       - Seed Sci. Technol. 7: 277-282, 1979.

6754 - BONNER, F.T., TURNER, B.J.: Rapid measurement of the moisture content of
       large seeds. - USDA Forest Serv. Tree Planters' Notes 31(3): 9-10, 1980.

*6755 - BORCHERT, R.: Differences in shoot growth patterns between juvenile and adult
       trees and their interpretation based on systems analysis of trees. - Acta
       Hort. 56: 123-130, 1976.

*6756 - BORCHERT, R.: Complete loss of stomatal functioning in aging leaves of tro-
       pical broadleafed trees. - Plant Physiol. 63 (Suppl.): 60, 1979.

6757 - BORKOVEC, V., PROCHÁZKA, S.: Vliv indolyloctové kyseliny (IAA) na transport
       abscisové kyseliny (ABA) v segmentech epikotylů hrachu (*Pisum sativum* L.).
       [Effect of indolylacetic acid upon the transport of abscisic acid in segments
       of pea (*Pisum sativum*) epicotyls.] - In: Dny Rostlinné Fyziologie II. Pp. 113-
       116. Vysoká Škola Zemědělská, Brno 1980. [In Czech, ab: E,R.]

*6758 - BORNMAN, C.H., HUBER, W.: *Nicotiana tabacum* callus studies. IX. Development
       in stressed explants. - Biochem. Physiol. Pflanz. 174: 345-356, 1979.

6759 - BORZAKIVS'KA, I.V., SHAĬTAN, I.M., KOLOMIĒTS', N.P., CHUPRINA, L.M.: Fosforni
       fraktsiĭ ta vodnyĭ rezhym u tkanynakh novykh gibrydiv persyka v zv'yazku z
       zymostiĭkistyu v protsesi aklimatyzatsiĭ. [Phosphorus fractions and water
       regime in tissues of new peach hybrids relative to their frost-resistance in
       the process of acclimatization.] - Ukr. bot. Zh. 37 (3): 48-51, 72, 1980. [In
       Ukr, ab: E.]

*6760 - BOT, G.P.A., DIXHOORN, J.J., van: Bond graphs and minicomputers in greenhouse
       climate control. - EPPO Bull. 9(3): 205-218, 1979.

*6761 - BOTHA, F.C.: The effect of drought conditions on water soluble proteins of
       two maize lines. - Z. Pflanzenphysiol. 95: 371-375, 1979.

*6762 - BOTHA, F.C., BOTHA, P.J.: The effect of water stress on the metabolism of
       two maize lines. II. Effects on the rate of protein synthesis and chlorophyll
       content. - Z. Pflanzenphysiol. 94: 179-183, 1979.

6763 - BOUAZIZ, E.: Tolérance à la salure de la Pomme de terre. - Physiol. vég. 18:
       11-17, 1980.

*6764 - BOUCAUD, J., BILLARD, J.-P.: Étude comparée des activités glutamate déshydro-
       génasique et glutamine synthétasique dans les racines et les parties aériennes
       d'un halophyte obligatoire: le *Suaeda maritima* var. *macrocarpa* et d'un glyco-
       phyte: le *Phaseolus vulgaris*, cultivés en présence de différentes concentra-
       tions en NaCl. - C.R. Acad. Sci. Paris, Sér. D 289: 599-602, 1979.

6765 - BOUZAIDI, A., EL AMANI, S.: Irrigation à l'eau salée de deux veriétés de
       Cotonnier dans les essais de plein champ. - Physiol. vég. 18: 35-44, 1980.

*6766 - BOWEN, G.D., CARTWRIGHT, B.: Mechanisms and models of plant nutrition. -
        In: RUSSELL, J.S., GREACEN, E.L. (ed.): Soil Factors in Crop Production in a
        Semi-Arid Environment. Pp. 197-223. University of Queensland Press, St. Lucia
        1977.

6767 - BOX, J.E.,Jr., WILKINSON, S.R., DAWSON, R.N., KOZACHYN, J.: Soil water effects
       on no-till corn production in strip and completely killed mulches. - Agron. J.
       72: 797-802, 1980.

6768 - BOYER, J.S.: Physiological adaptations to water stress. - In: TURNER, N.C.,
       KRAMER, P.J. (ed.): Adaptations of Plants to Water and High Temperature Stress.
       Pp. 443-444. John Wiley & Sons, New York - Chichester - Brisbane - Toronto
       1980.

6769 - BOYER, J.S., JOHNSON, R.R., SAUPE, S.G.: Afternoon water deficits and grain
       yields in old and new soybean cultivars. - Agron. J. 72: 981-986, 1980.

6770 - BOYER, J.S., MEYER, R.F.: Osmoregulation in plants during drought. - In:
       RAINS, D.W., VALENTINE, R.C., HOLLAENDER, A. (ed.): Genetic Engineering of
       Osmoregulation. Impact on Plant Productivity for Food, Chemicals, and Energy.
       Pp. 199-202. Plenum Press, New York - London 1980.

6771 - BOYER, J.S., YOUNIS, H.M.: Measurement of photosynthesis as a way to assess
       phytotoxicity. - J. environ. Sci. Health Pt.B 15(6): 1099-1104, 1980.

6772 - BRADFORD, J.M.: The penetration resistance in a soil with well-defined struc-
       tural units. - Soil Sci. Soc. Amer. J. 44: 601-606, 1980.

6773 - BRADFORD, J.M., BLANCHAR, R.W.: The effect of profile modification of a fra-
       giudalf on water extraction and growth by grain sorghum. - Soil Sci. Soc.
       Amer. J. 44: 374-378, 1980.

*6774 - BRADFORD, K.J., HSIAO, T.C.: Alterations in leaf angle and stomatal conduc-
        tance during waterlogging are independent of leaf water potential. - Plant
        Physiol. 63 (Suppl.): 88, 1979.

6775 - BRADFORD, K.J., YANG, S.F.: Xylem transport of 1-aminocyclopropane-1-carboxylic
       acid, an ethylene precursor, in waterlogged tomato plants. - Plant Physiol.
       65: 322-326, 1980.

6776 - BRADFORD, K.J., YANG, S.F.: Stress-induced ethylene production in the ethylene
       -requiring tomato mutant diageotropica. - Plant Physiol. 65: 327-330, 1980.

*6777 - BRAEKKE, F.H., KOZLOWSKI, T.T.: Effect of climatic and edaphic factors on
        radial stem growth of Pinus resinosa and Betula papyrifera in Northern Wiscon-
        sin. - Adv Front. Plant Sci. 30: 201-221, 1975.

*6778 - BRAMLAGE, W.J., LEOPOLD, A.C., SPECHT, J.E.: Imbibitional chilling sensitivity
        among soybean cultivars. - Crop Sci. 19: 811-814, 1979.

*6779 - BRANGEON, J., PRIOUL, J.L., REYSS, A.: Reversibility of light-induced adaptive
        responses in the rye-grass Lolium multiflorum. - In: COOMBS, J. (ed.): 4th
        International Congress on Photosynthesis. Pp. 53-54. UKISES, London 1977.

*6780 - BRAY, E., BRENNER, M.L., PARSONS, L.R.: Abscisic acid levels and stomatal
        resistance of red-osier dogwood during cold acclimation. - Plant Physiol.
        63 (Suppl.): 104, 1979.

*6781 - BRAYMAN, A., SCHAEDLE, M.: Stem photosynthesis in current-growth stems of
        Populus tremuloides Michx. - Plant Physiol. 63 (Suppl): 74, 1979.

6782 - BROWN, A.D., EDGLEY, M.: Osmoregulation in yeast. - In: RAINS, D.W., VALENTI-
       NE, R.C., HOLLAENDER, A.(ed.): Genetic Engineering of Osmoregulation. Impact
       on Plant Productivity for Food, Chemicals, and Energy. Pp. 75-90. Plenum Press,
       New York - London 1980.

*6783 - BROWN, A.R., PERKINS, H.F.: Single vs. double cropping under two levels of
        fertilization with and without irrigation. - Commun. Soil Sci. Plant Anal.
        10: 1279-1289, 1979.

6784 - BROWN, D.C.W., THORPE, T.A.: Changes in water potential and its components
       during shoot formation in tobacco callus. - Physiol. Plant. 49: 83-87, 1980.

*6785 - BROWN, F.A.,Jr., CHOW, C.S.: Interactions among beans in neighboring Faraday
        cages. - Experientia 33: 1316-1317, 1977.

6786 - BROWN, K.W., THOMAS, J.C.: The influence of water stress preconditioning on
       dark respiration. - Physiol. Plant. 49: 205-209, 1980.

6787 - BROWN, L.M., HELLEBUST, J.A.: The contribution of organic solutes to osmotic
       balance in some green and eustigmatophyte algae. - J. Phycol. 16: 265-270,
       1980.

6788 - BROWN, R.W., COLLINS, J.M.: A screen-caged thermocouple psychrometer and ca-
       libration chamber for measurements of plant and soil water potential. -
       Agron.J. 72: 851-854, 1980.

6789 - BRUCE, R.R., PALLAS, J.E.,Jr., HARPER, L.A., JONES, J.B.: Water and nutrient
       element regulation prescription in nonsoil media for greenhouse crop produc-
       tion. - Commun. Soil Sci. Plant Anal. 11: 677-698, 1980.

6790 - BRUCK, R.I., MANION, P.D.: Interacting environmental factors associated with
       the incidence of *Hypoxylon* canker on trembling aspen. - Can. J. Forest Res.
       10: 17-24, 1980.

*6791 - BUBL, C.E., RICHARDSON, D.G., MANSOUR, N.S.: Preharvest foliar desiccation
        and onion storage quality. - J. Amer. Soc. hort. Sci. 104: 773-777, 1979.

6792 - BUCKLEY, R.C., CORLETT, R.T., GRUBB, P.J.: Are the xeromorphic trees of tro-
       pical upper montane rain forests drought-resistant? - Biotropica 12: 124-136,
       1980.

*6793 - BUKOVAC, M.J., FLORE, J.A., BAKER, E.A.: Peach leaf surfaces: Changes in wet-
        tability, retention, cuticular permeability, and epicuticular wax chemistry
        during expansion with special reference to spray application. - J. Amer. Soc.
        hort. Sci. 104: 611-617, 1979.

6794 - BUKOVAC, M.J., WIDMOYER, F.B.: Observations on leaf characteristics of Afgha-
       nistan pine. - J. Amer. Soc. hort. Sci. 105: 293-297, 1980.

*6795 - BURCZYK, J., ZONTEK, I., SZURMAN, N.: Partition of various algal strains in
        two-polymer phase system of dextran and polyethylene glycol. - Bull. Acad.
        Pol. Sci. 26: 745-750, 1978.

6796 - BURESH, R.J., CASSELMAN, M.E., PATRICK, W.H.,Jr.: Nitrogen fixation in flooded
       soil systems, a review. - Adv. Agron. 33: 149-192, 1980.

*6797 - BURKE, M.J., STUSHNOFF, C.: Frost hardiness: A discussion of possible mole-
        cular causes of injury with particular reference to deep supercooling of
        water. - In: MUSSELL, H., STAPLES, R.(ed.): Stress Physiology in Crop Plants.
        Pp. 197-225. John Wiley & Sons, New York 1979.

*6798 - BURKINA, Z.S., GUSEINOVA, G.M.: Heavy-oxygen water uptake by *Zea mays* L.
roots in relation to the root functional state. - In: KUDREV, T., STOYANOV, I.,
GEORGIEVA, V. (ed.): Mineral Nutrition of Plants. Vol. II. Pp. 227-230.
Publishing House, Central Cooperative Union, Sofia 1979.

6799 - BURRIS, J.S.: Maintenance of soybean seed quality in storage as influenced
by moisture, temperature and genotype. - Iowa State J. Res. 54: 377-389, 1980.

*6800 - BURTON, G.W., HANNA, W.W., JOHNSON, J.C.,Jr., LEUCK, D.B., MONSON, W.G.,
POWELL, J.B., WELLS, H.D., WIDSTROM, N.W.: Pleiotropic effects of *tr* tricho-
meles gene in pearl millet on transpiration, forage quality, and pest resistance. - Crop Sci. 17: 613-616, 1977.

6801 - BUSBY, C.H., GUNNING, B.: Observations on pre-prophase bands of microtubules
in uniseriate hairs, stomatal complexes of sugar cane, and *Cyperus* root meristems. - Europ. J. Cell Biol. 21: 214-223, 1980.

6802 - BUTTERFASS, T.: The continuity of plastids and the differentiation of plastid
populations. - In: REINERT, J. (ed.): Chloroplasts. Pp. 29-44. Springer-Verlag,
Berlin - Heidelberg - New York 1980.

6803 - BUVAT, R.: Restructurations cytoplasmiques au cours de l'hydration germinative
des ébauches de racines du caryopse de l'Orge (*Hordeum sativum*). Activités
phosphatasiques acides et réactivation des dictyosomes. - C.R. Acad. Sci.
Paris, Sér. D 290: 1031-1034, 1980.

*6804 - BUZÁS, I., SERES, I.: A nitrogéntrágyázás és öntözés hatása a cukorrépára.
[The effect of nitrogen fertilization and watering on the sugar beet.] -
Cukoripar 28: 121-124, 1975. [In Hung, ab: E,G,R.]

6805 - BYKOV, O.D.: Issledovaniya po fotosintezu v svyazi s zadachami selektsii kak
nauki. [Photosynthesis study in connection with tasks of breeding as a science.] - Sel'skokhoz. Biol. 15: 334-341, 1980. [In R, ab: E.]

*6806 - BYKOV, O.D., GALKIN, V.I., ZHITLOVA, N.A., KOSHKIN, V.A.: Vliyanie temperatu-
ry na izmenenie intensivnosti fotosinteza i dykhaniya diploidov i poliploidov
kartofelya. [Effect of temperature on changes in photosynthetic and respiration rates of potato diploids and polyploids.] - In: NASYROV, Yu.S. (ed.):
Genetika Fotosinteza. Pp. 241-249. Donish, Dushanbe 1977. [In R.]

6807 - CABELGUENNE, M.: Etude technique et économique de la valorisation de l'irrigation du maïs (région Toulousaine). - Ann. agron. 31: 319-336, 1980.

6808 - CACERES, J.R., ROJAS GARCIDUEÑAS, M.: Response of drought-resistant and
drought-susceptible maize cultivars to chlormequat application. - Turrialba
30: 25-28, 1980.

6809 - CAIAZZA, N.A.,Jr., QUINN, J.A.: Leaf morphology in *Arenaria patula* and *Lonicera japonica* along a pollution gradient. - Bull. Torrey bot. Club 107: 9-18,
1980.

*6810 - CAIN, D.W., ANDERSEN, R.L.: Temperature and moisture effects on wood injury
of cold-stressed 'Siberian C' and 'Redhaven' peaches. - HortScience 14: 518-
519, 1979.

*6811 - CALDER, I.R.: The measurement of water losses from a forested area using a
"natural" lysimeter. - J. Hydrol. 30: 311-325, 1976.

6812 - CALDERWOOD, D.L., BOLLICH, C.N., SCOTT, J.E.: Field drying of rough rice:
Effect on grain yield, milling quality, and energy saved. - Agron. J. 72:
649-653, 1980.

*6813 - CALLAGHAN, P.T., JOLLEY, K.W., LELIEVRE, J.: Diffusion of water in the endo-
sperm tissue of wheat grains as studied by pulsed field gradient nuclear mag-
netic resonance. - Biophys. J. 28: 133-142, 1979.

6814 - CAMPBELL, J.A., MILLETTE, J.A., ROY, M.: An inexpensive instrument for measuring soil water table levels. - Can. J. Soil Sci. 60: 575-577, 1980.

6815 - CANNELL, R.Q., BELFORD, R.K.: Effects of waterlogging at different stages of
development on the growth and yield of winter oilseed rape (*Brassica napus* L.).
- J. Sci. Food Agr. 31: 963-965, 1980.

6816 - CANNELL, R.Q., BELFORD, R.K., GALES, K., DENNIS, C.W.: A lysimeter system
used to study the effect of transient waterlogging on crop growth and yield. -
J. Sci. Food Agr. 31: 105-116, 1980.

6817 - CANNELL, R.Q., BELFORD, R.K., GALES, K., DENNIS, C.W., PREW, R.D.: Effects of
waterlogging at different stages of development on the growth and yield of
winter wheat. - J. Sci. Food Agr. 31: 117-132, 1980.

*6818 - CANVIN, D.T.: Photorespiration and the effect of oxygen on photosynthesis. -
In: SIEGELMAN, H.W., HIND, G. (ed.): Photosynthetic Carbon Assimilation. Pp.
61-76. Plenum Press, New York - London 1978.

6819 - CARBON, B.A., BARTLE, G.A., MURRAY, A.M., MACPHERSON, D.K.: The distribution
of root length, and the limits to flow of soil water to roots in a dry scle-
rophyll forest. - Forest Sci. 26: 656-664, 1980.

6820 - CARCELLER, M., FRASCHINA, A.: The free proline content of water stressed
maize roots. - Z. Pflanzenphysiol. 100: 43-49, 1980.

6821 - CARCELLER, M., FRASCHINA, A.: Acumulación de prolina libre en plántulas de
maíz y su relación con la resistencia a la sequía. [Accumulation of free proline in maize in relation to drought resistance.] - Turrialba 30: 231-233,
1980. [In Span, ab: E.]

*6822 - CARCELLER, M., SORIANO, A.: Acción de las citocininas sobre el crecimiento de
plántulas de trigo sometidas a sequía. [Effects of cytokinins on growth of
wheat plants affected by drought.] - Turrialba 27: 293-298, 1977. [In Span,
ab: E.]

6823 - CARLSON, R.W., BAZZAZ, F.A.: The effects of elevated CO_2 concentrations on
growth, photosynthesis, transpiration, and water use efficiency of plants. -
In: SINGH, J.J., DEEPAK, A. (ed.): Environmental and Climatic Impact of Coal
Utilization. Pp. 609-623. Academic Press, New York 1980.

*6824 - CARLSSON, H.: Förgroning och bevattning av matpotatis. [Presprouting and
irrigation of table potatoes.] - Lantbrukshögsk. Medd. Ser. A 235: 1-25,
1975. [In Swed, ab: E.]

*6825 - CARMI, A., KOLLER, D.: Regulation of photosynthetic activity in the primary
leaves of bean (*Phaseolus vulgaris* L.) by materials moving in the water-
-conducting system. - Plant Physiol. 64: 285-288, 1979.

6826 - CARR, D.J., CARR, S.G.M.: *Eucalyptus* stomata with occluded anterior chambers.
- Protoplasma 104: 239-251, 1980.

6827 - CARR, D.J., CARR, S.G.M., JAHNKE, R.: Intercellular strands associated with
stomata: Stomatal pectic strands. - Protoplasma 102: 177-182, 1980.

*6828 - CARR, W.W., BALLARD, T.M.: Effects of fertilizer salt concentration on viabi-
lity of seed and *Rhizobium* used for hydroseeding. - Can. J. Bot. 57: 701-704,
1979.

6829 - CARTER, J.N., JENSEN, M.E., TRAVELLER, D.J.: Effect of mid- to late-season
       water stress on sugarbeet growth and yield. - Agron. J. 72: 806-815, 1980.

*6830 - CARTER, J.N., PAIR, C.H., WESTERMANN, D.T.: Effect of irrigation method and
       late season nitrate-nitrogen concentration on sucrose production by sugarbeets.
       - J. Amer. Soc. Sugar Beet Technol. 18: 332-342, 1975.

6831 - CARTER, J.V., BRADEN, M.: Lethal freeze-dehydration injury of dogwood stem
       tissue does not change the activation energy of water permeability. - Plant
       Physiol. 65: 499-501, 1980.

*6832 - CARY, J.W.: Relations between CO_2 exchange rate, CO_2 compensation, and meso-
       phyll resistance from a simple field method. - Crop Sci. 17: 453-456, 1977.

6833 - ČATSKÝ, J., TICHÁ, I.: Ontogenetické změny závislosti vodivostí pro přenos CO_2
       na kvantové ozářenosti listu. [Ontogenetic changes in the dependence of con-
       ductances for CO_2 transfer on leaf quantum irradiance.] - In: Dny Rostlinné
       Fyziologie II. Pp. 118-121. Vysoká Škola Zemědělská, Brno 1980. [In Czech,
       ab: E,R.]

6834 - CAVINESS, C.E., THOMAS, J.D.: Yield reduction from defoliation of irrigated
       and non-irrigated soybeans. - Agron. J. 72: 977-980, 1980.

*6835 - CERDÁ, A., FERNÁNDEZ, F.G., CARO, M., GUILLÉN, M.G.: Growth and mineral compo-
       sition of two lemon varieties irrigated with saline waters. - Agrochimica 23:
       387-396, 1979.

6836 - ČERMÁK, J., HUZULÁK, J., PENKA, M.: Water potential and sap flow rate in
       adult trees with moist and dry soil as used for the assessment of root system
       depth. - Biol. Plant. 22: 34-41, 1980.

*6837 - ČERMÁK, J., KUČERA, J., PRAX, A., ŽÍDEK, V.: Tok vody v systému půda-rostlina
       -atmosféra na příkladu břízy v lesním porostu. [Water flow in the continuum
       soil-plant-atmosphere on the example of birch in forest stand.] - In: Bilancia
       Energie a Vody v Poľných a Lesných Ekosystémoch. Pp. 135-149. Vysoká Škola
       Poľnohospodárská, Nitra 1979. [In Czech.]

*6838 - CERNUSCA, A. (ed.): Alpine Grasheide Hohe Tauern. Ergebnisse der Ökosystem-
       studie 1976. - Universitätsverlag Wagner, Innsbruck 1977.

6839 - CEULEMANS, R., IMPENS, I.: Leaf gas exchange processes and related characte-
       ristics of seven poplar clones under laboratory conditions. - Can. J. Forest
       Res. 10: 429-435, 1980.

6840 - CEULEMANS, R., IMPENS, I., GABRIËLS, R.: Comparative study of photosynthesis,
       transpiration, diffusion resistances and water-use efficiency of two azalea
       cultivars. - Scientia Hort. 13: 283-288, 1980.

6841 - CEULEMANS, R., IMPENS, I., HEBRANT, F., MOERMANS, R.: Evaluation of field
       productivity for several poplar clones based on their gas exchange variables
       determined under laboratory conditions. - Photosynthetica 14: 355-362, 1980.

*6842 - CHABOT, B.F.: Metabolic and enzymatic adaptations to low temperature. - In:
       UNDERWOOD, L.S., TIESZEN, L.L., CALLAHAN, A.B., FOLK, G.E. (ed.): Comparative
       Mechanisms of Cold Adaptation. Pp. 283-301. Academic Press, New York 1979.

6843 - CHAMEL, A.: Pénétration du cuivre à travers des cuticules isolées de feuilles
       de Poirier. - Physiol. vég. 18: 313-323, 1980.

*6844 - CHANDRA, P.: Leaf epidermis in some species of *Asplenium* L. - Proc. Ind. Acad.
       Sci. 88 B: 269-275, 1979.

6845 - CHANG, M., AGUILAR, G.J.R.: Effects of climate and soil on the radial growth
of loblolly pine (*Pinus taeda* L.) in a humid environment of southeastern U.S.A.
- Forest. Ecol. Manage. 3: 141-150, 1980.

*6846 - CHASE, R.L., APPLEBY, A.P.: Effects of humidity and moisture stress on glypho-
sate control of *Cyperus rotundus* L. - Weed Res. 19: 241-246, 1979.

*6847 - CHEEMA, S.S., KUNDRA, H.: Timing last irrigation to wheat. - Ind. J. Agron.
22: 172-173, 1977.

6848 - CHELLAPPAN, K.P., SEENI, S., GNANAM, A.: Photosynthetic studies with mesophyll
protoplasts from *Notonia grandiflora,* a crassulacean acid metabolism plant. -
Physiol. Plant. 48: 403-410, 1980.

6849 - CHEN, H.H., LI, P.H.: Biochemical changes in tuber-bearing *Solanum* species in
relation to frost hardiness during cold acclimation. - Plant Physiol. 66:
414-421, 1980.

6850 - CHETAL, S., WAGLE, D.S., NAINAWATEE, H.S.: Phospholipid changes in wheat and
barley leaves under water stress. - Phytochemistry 19: 1393-1395, 1980.

6851 - CHOTENA, M., MAKUS, D.J., SIMPSON, W.R.: Effect of water stress on production
and quality of sweet corn seed. - J. Amer. Soc. hort. Sci. 105: 289-293, 1980.

*6852 - CHOUDHURI, G.N., SHARMA, B.D.: Response of intraspecific variants of a fa-
cultative halophyte to cationic treatments. - Geobios 6: 64-66, 1979.

6853 - CHOUDHURY, P.N., KUMAR, V.: The sensitivity of growth and yield of dwarf
wheat to water stress at three growth stages. - Irrig. Sci. 1: 223-231, 1980.

*6854 - CHOW, W.S., THORNE, S.W., BOARDMAN, N.K.: Formation of the proton gradient
across the chloroplast thylakoid membrane in relation to ATP synthesis. -
In: DUTTON, P.L., LEIGH, J.S., SCARPA, A. (ed.): Frontiers of Biological
Energetics: Electrons to Tissues. Vol. 1. Pp. 287-296. Academic Press, New
York - San Francisco - London 1978.

6855 - CHUNYUN, G., GUOHUAI, W., JUNTIAN, Z.: [The preliminary investigation on
heterosis and early prediction in heterosis selection of hybrids of rapaseed
(*Brassica napus*).] - Acta gen. Sinica 7: 55-63, 1980. [In Sin, ab: E.]

6856 - ČIAMPOROVÁ, M.: Ultrastructure of cortical cells of maize root under water
stress. - Biol. Plant. 22: 444-449, 1980.

6857 - ČIAMPOROVÁ, M.: Štruktúrne zmeny koreňových buniek pri nedostatku vody.
[Structural changes of the root cells under water deficit.] - In: Dny Rost-
linné Fyziologie II. Pp. 133-137. Vysoká Škola Zemědělská, Brno 1980. [In
Slov, ab: E.]

6858 - CIPRA, J.E., NOGUERAPENA, N.E., BRYSON, M.C., LUEKING, M.A.: Forage production
estimates for irrigated meadows from Landsat data. - Agron. J. 72: 793-796,
1980.

6859 - CLARK, R.J., MENARY, R.C.: The effect of irrigation and nitrogen on the yield
and composition of peppermint oil (*Mentha piperita* L.). - Aust. J. agr. Res.
31: 489-498, 1980.

*6860 - CLARKE, A.L., RUSSELL, J.S.: Crop sequential practices. - In: RUSSELL, J.S.,
GREACEN, E.L. (ed.): Soil Factors in Crop Production in a Semi-Arid Environment. Pp. 279-300. University of Queensland Press, St. Lucia 1977.

6861 - CLARKE, J.M.: Measurement of relative water uptake rates of wheat seeds using
agar media. - Can. J. Plant Sci. 60: 1035-1038, 1980.

6862 - CLARKE, R.W., JENNINGS, D.H., COGGINS, C.R.: Growth of *Serpula lacrimans* in relation to water potential of substrate. - Trans. Brit. mycol. Soc. 75: 271-280, 1980.

6863 - CLARKSON, D.T., HANSON, J.B.: The mineral nutrition of higher plants. - Annu. Rev. Plant Physiol. 31: 239-298, 1980.

6864 - CLELAND, R.E.: Auxin and H^+-excretion: the state of our knowledge. - In: SKOOG, F. (ed.): Plant Growth Substances 1979. Pp. 71-78. Springer-Verlag, Berlin - Heidelberg - New York 1980.

*6865 - CLEMMENS, A.J.: Verification of the zero-inertia model for border irrigation. - Trans. ASAE 22: 1306-1309, 1979.

6866 - CLOUGH, J.M., ALBERTE, R.S., TEERI, J.A.: Photosynthetic adaptation of *Solanum dulcamara* L. to sun and shade environments. III. Characterization of genotypes with differing photosynthetic performance. - Oecologia 44: 221-225, 1980.

*6867 - CLUFF, C.B.: The use of the compartmented reservoir in water harvesting agri-systems. - In: GOODIN, J.R., NORTHINGTON, D.K. (ed.): Arid Land Plant Resources. Pp. 482-500. Texas Technical University, Lubbock 1979.

*6868 - COCHRANE, M.P., DUFFUS, C.M.: Morphology and ultrastructure of immature cereal grains in relation to transport. - Ann. Bot. 44: 67-72, 1979.

6869 - COELHO, D.T., DALE, R.F.: An energy-crop growth variable and temperature function for predicting corn growth and development: Planting to silking. - Agron. J. 72: 503-510, 1980.

6870 - COGGINS, C.R., JENNINGS, D.H., CLARKE, R.W.: Tear or drop formation by mycelium of *Serpula lacrimans*. - Trans. Brit. mycol. Soc. 75: 63-67, 1980.

*6871 - COKE, L., SIONIT, N.: Stomatal behavior and gas exchange in leaves of cassava (*Manihot esculenta*) treated with abscisic acid and vomifoliol. - Plant Physiol. 63 (Suppl.): 121, 1979.

6872 - COLLINS, J.C., MORGAN, M.: The influence of temperature on the abscisic acid stimulated water flow from excised maize roots. - New Phytol. 84: 19-26, 1980.

6873 - COLLINS, N.J., CALLAGHAN, T.V.: Predicted patterns of photosynthetic production in maritime antarctic mosses. - Ann. Bot. 45: 601-620, 1980.

6874 - CONSTABLE, G.A., HEARN, A.B.: Irrigation for crops in a sub-humid environment. I. The effect of irrigation on the growth and yield of soybeans. - Irrig. Sci. 2: 1-12, 1980.

6875 - CONSTABLE, G.A., RAWSON, H.M.: Photosynthesis, respiration and transpiration of cotton fruit. - Photosynthetica 14: 557-563, 1980.

6876 - CONSTABLE, G.A., RAWSON, H.M.: Effect of leaf position, expansion and age on photosynthesis, transpiration and water use efficiency of cotton. - Aust. J. Plant Physiol. 7: 89-100, 1980.

6877 - COOKE, J.R., RAND, R.H.: Diffusion resistance models. - In: HESKETH, J.D., JONES, J.W. (ed.): Predicting Photosynthesis for Ecosystem Models. Vol. I. Pp. 93-121. CRC Press, Boca Raton 1980.

6878 - COOKE, R.J., ROBERTS, K., DAVIES, D.D.: Model for stress-induced protein degradation in *Lemna minor*. - Plant Physiol. 66: 1119-1122, 1980.

6879 - COOKE, T.J., PAOLILLO, D.J.,Jr.: Dark growth and the associated phenomenon
       of age-dependent photoresponses in fern gametophytes. - Ann. Bot. 45: 693-702,
       1980.

*6880 - COOLBEAR, P., GRIERSON, D.: Studies on the changes in the major nucleic acid
       components of tomato seeds (*Lycopersicon esculentum* Mill.) resulting from
       osmotic presowing treatment. - J. exp. Bot. 30: 1153-1162, 1979.

6881 - COOLBEAR, P., GRIERSON, D., HEYDECKER, W.: Osmotic pre-sowing treatments and
       nucleic acid accumulation in tomato seeds (*Lycopersicon lycopersicum*). -
       Seed Sci. Technol. 8: 289-303, 1980.

6882 - COOPER, J.L.: The effect of nitrogen fertilizer and irrigation frequency on
       a semi-dwarf wheat in south-east Australia. 1. Growth and Yield. - Aust. J.
       exp. Agr. anim. Husb. 20: 359-364, 1980.

6883 - COOPER, J.L.: The effect of nitrogen fertilizer and irrigation frequency on
       a semi-dwarf wheat in south-east Australia. 2. Water use. - Aust. J. exp.
       Agr. anim. Husb. 20: 365-369, 1980.

6884 - COOPER, R.M., WOOD, R.K.S.: Cell wall degrading enzymes of vascular wilt
       fungi. III. Possible involvement of endo-pectin lyase in *Verticillium* wilt
       of tomato. - Physiol. Plant Pathol. 16: 285-300, 1980.

6885 - CORNELIUS, R.: Synergistische Wirkungen von Auftausalzen und SO_2 auf die
       Nettophotosynthese von Gehölzen. - Angew. Bot. 54: 329-335, 1980.

6886 - COUCHAT, P., LASCEVE, G.: Tritiated water vapour exchange method for the
       evaluation of whole plant diffusion resistance. - J. exp. Bot. 31: 1217-1222,
       1980.

6887 - COUCHAT, P., LASCÈVE, G.: Intervention de l'oxygène atmosphérique sur la
       réponse hydrique du Tournesol à une variation brutale de température. -
       C.R. Acad. Sci. Paris Sér. D 290: 271-274, 1980.

6888 - COUDRET, A., FERRON, F., GAUDILLÈRE, J.-P., COSTES, C.: Action comparée des
       antitranspirants sur le mouvement des stomates, les échanges de CO_2 et la
       production de matière sèche chez *Plantago lanceolata* L. et *Plantago maritima*
       L. - Physiol. vég. 18: 631-643, 1980.

6889 - COUDRET, A., LOUGUET, P.: Étude comparée de l'action du NaCl sur les mouve-
       ments stomatiques de *Plantago maritima* L. var. Graminaea et de *Plantago
       lanceolata* L. - Physiol. vég. 18: 55-68, 1980.

*6890 - COUPLAND, D., CASELEY, J.C.: Presence of ^{14}C activity in root exudates and
       guttation fluid from *Agropyron repens* treated with ^{14}C-labelled glyphosate. -
       New Phytol. 83: 17-22, 1979.

6891 - COUTTS, M.P.: Control of water loss by actively growing Sitka spruce seedlings
       after transplanting. - J. exp. Bot. 31: 1587-1597, 1980.

*6892 - COWAN, D.A., GREEN, T.G.A., WILSON, A.T.: Lichen metabolism. 1. The use of
       tritium labelled water in studies of anhydrobiotic metabolism in *Ramalina
       celastri* and *Peltigera polydactyla*. - New Phytol. 82: 489-503, 1979.

6893 - COWLING, J.E., KEDROWSKI, R.A.: Winter water relations of native and intro-
       duced evergreens in interior Alaska. - Can. J. Bot. 58: 94-99, 1980.

6894 - CROWE, F.J., HALL, D.H.: Soil temperature and moisture effects on sclerotium
       germination and infection of onion seedlings by *Sclerotium cepivorum*. -
       Phytopathology 70: 74-78, 1980.

6895 - CROWLEY, R.H., BUCHANAN, G.A.: Responses of *Ipomoea* spp. and smallflower
       morningglory (*Jacquemontia tamnifolia*) to temperature and osmotic stresses. -
       Weed Sci. 28: 76-82, 1980.

6896 - CRUIZIAT, P., TYREE, M.T., BODET, C., LoGULLO, M.A.: The kinetics of rehydra-
       tion of detached sunflower leaves following substantial water loss. - New
       Phytol. 84: 293-306, 1980.

6897 - CSONKA, L.N.: The role of L-proline in response to osmotic stress in *Salmo-
       nella typhimurium*: Selection of mutants with increased osmotolerance as
       strains which over-produce L-proline. - In: RAINS, D.W., VALENTINE, R.C.,
       HOLLAENDER, A. (ed.): Genetic Engineering of Osmoregulation. Impact on Plant
       Productivity for Food, Chemicals, and Energy. Pp. 35-52. Plenum Press, New
       York - London 1980.

*6898 - CUMING, A.C., LANE, B.G.: Protein synthesis in imbibing wheat embryos. -
       Europ. J. Biochem. 99: 217-224, 1979.

6899 - CUTLER, J.M., SHAHAN, K.W., STEPONKUS, P.L.: Alteration of the internal water
       relations of rice in response to drought hardening. - Crop Sci. 20: 307-310,
       1980.

6900 - CUTLER, J.M., SHAHAN, K.W., STEPONKUS, P.L.: Dynamics of osmotic adjustment
       in rice. - Crop Sci. 20: 310-314, 1980.

6901 - CUTLER, J.M., SHAHAN, K.W., STEPONKUS, P.L.: Influence of water deficits and
       osmotic adjustment on leaf elongation in rice. - Crop Sci. 20: 314-318, 1980.

*6902 - DAHIYA, I.S., HAJRASULIHA, S., LAMBA, P.S.: A quick method of soil moisture
       determination. - Commun. Soil Sci. Plant Anal. 10: 795-805, 1979.

*6903 - DAMISCH, W., WIBERG, A.: Ergebnisse über den Einfluss der Temperatur auf
       Stoffzuwachs und Ertragsbildung bei Sommergerste. - Arch. Acker- Pflanzenbau
       Bodenk. 21: 485-494, 1977.

6904 - DANIEL, V., GAFF, D.F.: Sulphydryl and disulphide levels in protein fractions
       from hydrated and dry leaves of resurrection plants. - Ann. Bot. 45: 163-171,
       1980.

6905 - DANIEL, V., GAFF, D.F.: Dessication-induced changes in the protein, comple-
       ment of soluble extracts from leaves of resurrection plants and related
       desiccation-sensitive species. - Ann. Bot. 45: 173-181, 1980.

*6906 - DARBY, R.J., SALTER, P.J., WHITLOCK, A.J.: Effects of osmotic treatment and
       pre-germination of celery seeds on seedling emergence. - Exp. Hort. 31:
       10-20, 1979.

*6907 - DAS, V.S.R., RAO, I.M., SWAMY, P.M.: Antitranspirant activity of morphactin
       on cotton plants. - Ind. J. exp. Biol. 15: 642-644, 1977.

*6908 - DAS, V.S.R., VEERANJANEYULU, K.: Leaf gas exchange characteristics of twenty-
       six tropical weed and crop plants. - Plant Physiol. 63 (Suppl.): 141, 1979.

6909 - DAVENPORT, T.L., MORGAN, P.W., JORDAN, W.R.: Reduction of auxin transport
       capacity with age and internal water deficits in cotton petioles. - Plant
       Physiol. 65: 1023-1025, 1980.

*6910 - DAVIES, F.S., LAKSO, A.N.: Diurnal and seasonal changes in leaf water poten-
       tial components and elastic properties in response to water stress in apple
       trees. - Physiol. Plant. 46: 109-114, 1979.

*6911 - DAVIES, F.S., TERAMURA, A.H., BUCHANAN, D.W.: Yield, stomatal resistance, xylem pressure potential, and feeder root density in three rabbiteye blueberry cultivars. - HortScience 14: 725-726, 1979.

6912 - DAVIES, W.J., MANSFIELD, T.A., WELLBURN, A.R.: A role for abscisic acid in drought endurance and drought avoidance. - In: SKOOG, F. (ed.): Plant Growth Substances 1979. Pp. 242-253. Springer-Verlag, Berlin - Heidelberg - New York 1980.

*6913 - DAVIS, E.A.: Root system of shrub live oak in relation to water yield by chaparral. - Hydrol. Water Resour. Arizona Southwest 7: 241-248, 1977.

*6914 - DÉCAMPS, O.: Caractères stomatiques des *Ranunculacées*. - Bull. Soc. Histoire natur. Toulouse 114: 429-446, 1978.

*6915 - DeJONG, T.M., DRAKE, B.G.: Comparative laboratory and field gas exchange responses of C_3 and C_4 tidal marsh species. - Plant Physiol. 63 (Suppl.): 63, 1979.

*6916 - DEKOV, I., KUDREV, T., PETROVA, L.: Influence of magnesium deficiency and water stress on ultrastructural changes of chloroplasts and some parameters of water regime in maize plants. - In: KUDREV, T., STOYANOV, I., GEORGIEVA, V. (ed.): Mineral Nutrition of Plants. Vol. I. Pp. 133-138. Publishing House, Central Cooperative Union, Sofia 1979.

6917 - De MORAES, J.A.P.V.: CO_2-gas exchange parameters of palm seedlings (*Washingtonia filifera* and *Serenoa repens*). - Acta oecol. - Oecol. Plant. 1: 299-305, 1980.

6918 - DENGLER, N.G.: Comparative histological basis of sun and shade leaf dimorphism in *Helianthus annuus*. - Can. J. Bot. 58: 717-730, 1980.

6919 - DENGLER, N.G.: The histological basis of leaf dimorphism in *Selaginella martensii*. - Can. J. Bot. 58: 1225-1234, 1980.

*6920 - DENMEAD, O.T.: Temperate cereals. - In: MONTEITH, J.L. (ed.): Vegetation and Atmosphere. Vol. 2. Case Studies. Pp. 1-31. Academic Press, London - New York - San Francisco 1976.

6921 - DERCO, M.: Synergizmus závlahy a hnojív v závislosti od ekologických podmienok. [The synergism of irrigation and fertilizers as depending on ecological conditions.] - Rost. Výroba (Praha) 26: 563-573, 1980. [In Slov, ab: R,E,G.]

6922 - DERCO, M., BARTA, V.: Vplyv ošetrovania porastu zavlažovanej kukurice na úrodu zrna a na hospodárenie s pôdnou vodou. [The effect of the treatment of an irrigated maize stand on grain yields and on soil moisture regime.] - Rost. Výroba (Praha) 26: 123-132, 1980. [In Slov, ab: R,E,G.]

*6923 - DEREUDDRE, J.: Etude comparative du comportement des bourgeons d'arbres en vie ralentie, pendant un refroidissement graduel des rameaux. - Bull. Soc. bot. France, Lett. bot. 126: 399-412, 1979.

6924 - DE STIGTER, H.C.M.: Water balance of cut and intact "Sonia" rose plants. - Z. Pflanzenphysiol. 99: 131-140, 1980.

*6925 - DEVANATHAN, M.A.V.: Photosynthetic productivity in natural environments. - In: COOMBS, J. (ed.): 4th International Congress on Photosynthesis. P. 93. UKISES, London 1977.

6926 - DIAMANTOGLOU, S., MELETIOU-CHRISTOU, M.-S.: Kohlenhydratgehalte und osmotische Verhältnisse bei Blättern und Rinden von *Pistacia lenticus, Pistacia terebinthus* und *Pistacia vera* im Jahresgang. - Flora 169: 168-176, 1980.

*6927 - DiCAMELLI, C.A., OUTLAW, W.H.: Isolation and characterization of guard cell protoplast phosphoenolpyruvate carboxylase. - Plant Physiol. 63 (Suppl.): 60, 1979.

*6928 - DICKSON, R.E.: Xylem translocation of amino acids from roots to shoots in cottonwood plants. - Can. J. Forest Res. 9: 374-378, 1979.

6929 - DIKIÏ, S.P., ANIKEENKO, A.P., FONINA, T.I.: Izmenenie biokhimicheskogo sostava plodov baklazhana pod vliyaniem orosheniya. [Changes in the aubergine fruit biochemical composition under irrigation effect.] - Fiziol. Biokhim. kul't. Rast. 12: 170-174, 1980. [In R, ab: E.]

*6930 - DIMITROV, Kh., DIMITROVA, M.: Vliyanie na mineralnoto khranene i vodosnabdyavaneto v"rkhu fotosintezata i rastezha na topolite. [Effect of mineral nutrition and water supply on poplar tree photosynthesis and growth.] - Fiziol. Rast. (Sofia) 2 (3): 43-51, 1976. [In Bulg, ab: E.]

6931 - DIXON, R.K., WRIGHT, G.M., BEHRNS, G.T., TESKEY, R.O., HINCKLEY, T.M.: Water deficits and root growth of ectomycorrhizal white oak seedlings. - Can. J. Forest Res. 10: 545-548, 1980.

*6932 - DOEHLERT, D.C., KU, M.S.B., EDWARDS, G.E.: Dependence of the post-illumination burst of CO_2 on temperature, light, CO_2 and O_2 concentration in wheat (*Triticum aestivum*). - Physiol. Plant. 46: 299-306, 1979.

*6933 - DOÏKOV, K., NENOVA, L., VITKOV, M.: Vliyanie na napoyavaneto i toreneto v"rkhu ikonomicheskata efektivnost na tsarevitsata za z"rno. [Influence of irrigation and fertilization on economic effectiveness of production of corn for grain.] - Rasteniev. Nauki 16 (4): 76-83, 1979. [In Bulg, ab: E,R.]

*6934 - DOLEY, D.: Parthenium weed (*Parthenium hysterophorus* L.): Gas exchange characteristics as a basis for prediction of its geographical distribution. - Aust. J. agr. Res. 28: 449-460, 1977.

*6935 - DOLGOPOLOVA, L.N., LAKHANOV, A.P., CHERNEN'KAYA, R.F.: Aminokislotnyĭ sostav belka list'ev gorokha v svyazi s ustoĭchivost'yu k pochvenoĭ zasukhe.[Amino acid composition of pea leaf protein regarding resistance to soil drought.] - Fiziol. Biokhim. kul't. Rast. 11: 158-163, 1979. [In R, ab: E.]

6936 - DOLPH, G.E., DILCHER, D.L.: Variation in leaf size with respect to climate in Costa Rica. - Biotropica 12: 91-99, 1980.

*6937 - DOMERGUE, F., de CORMIS, L., LOUGUET, P.: Influence d'une pollution de l'air par l'ozone sur le degré d'overture stomatique du *Pelargonium* X *hortorum*. - C.R. Acad. Sci. Paris, Sér. D 288: 1541-1544, 1979.

*6938 - DOMINY, P.J., BAKER, N.R.: Effect of high salinity on primary photochemistry and energy transfer in pea thylakoids. - Plant Physiol. 61 (Suppl): 88, 1978.

*6939 - DOMINY, P.J., BAKER, N.R.: Effect of salinity on PSII primary photochemistry of pea and sea beet. - Plant Physiol. 63 (Suppl.): 89, 1979.

6940 - DOMINY, P.J., BAKER, N.R.: Salinity and *in vitro* ageing effects on primary photosynthetic processes of thylakoids isolated from *Pisum sativum* and *Spinacia oleracea*. - J. exp. Bot. 31: 59-74, 1980.

*6941 - DONALDSON, E., NILAN, R.A., KONZAK, C.F.: Minimum gamma-radiation exposure and oxygen concentration to produce post-irradiation oxygen-enhancement of damage in barley seeds. - Environ. exp. Bot. 19: 165-173, 1979.

6942 - DONKIN, M.E., MARTIN, E.S.: Studies on the properties of carboxylating enzymes in the epidermis of *Commelina communis*. - J. exp. Bot. 31: 357-363, 1980.

6943 - DONKIN, M.E., MARTIN, E.S.: Changes in starch and glucose levels in the epidermis of *Commelina communis* in relation to stomatal movements. - Plant Cell
Environ. 3: 409-414, 1980.

6944 - DOOHAN, M.E., PALEVITZ, B.A.: Microtubules and coated vesicles in guard-cell
protoplasts of *Allium cepa* L. - Planta 149: 389-401, 1980.

*6945 - DORAISWAMY, P.C., HODGES, T., PHINNEY, D.E.: Crop yield literature review for
AgRISTARS crops. Corn, soybeans, wheat, barley, sorghum, rice, cotton, and
sunflowers. - In: Tech. Rep. No. SR-L9-00405; JSC-16320. Lockheed Eng. Manage.
Serv. Co., Houston 1979.

6946 - DÖRFFLING, K., TIETZ, D., STREICH, J., LUDEWIG, M.: Studies on the role of
abscisic acid in stomatal movements. - In: SKOOG, F. (ed.): Plant Growth
Substances 1979. Pp. 274-285. Springer-Verlag, Berlin - Heidelberg - New York
1980.

*6947 - DORNHOFF, G.M., SHIBLES, R.: Leaf morphology and anatomy in relation to
CO_2-exchange rate of soybean leaves. - Crop Sci. 16: 377-381, 1976.

6948 - DORTENZIO, W.A., NORRIS, R.F.: The influence of soil moisture on the foliar
activity of diclofop. - Weed Sci. 28: 534-539, 1980.

6949 - DOSS, B.D., TURNER, J.L., EVANS, C.E.: Irrigation methods and in-row chiseling
for tomato production. - J. Amer. Soc. hort. Sci. 105: 611-614, 1980.

6950 - DOTY, C.W.: Crop water supplied by controlled and reversible drainage. -
Trans. ASAE 23: 1122-1126, 1130, 1980.

6951 - DOWDELL, R.J., WEBSTER, C.P.: A lysimeter study using nitrogen-15 on the
uptake of fertilizer nitrogen by perennial ryegrass swards and losses by
leaching. - J. Soil Sci. 31: 65-75, 1980.

6952 - DRACUP, J.A., LEE, K.S., PAULSON, E.G.,Jr.: On the definition of droughts. -
Water Resour. Res. 16: 297-302, 1980.

6953 - DRAKE, R.J., PEPPER, I.L., JOHNSON, G.V., KNEEBONE, W.R.: Design and testing
of a new microlysimeter for leaching studies. - Agron. J. 72: 397-398, 1980.

*6954 - DRAKE, S.R., NELSON, J.W.: A comparison of three methods of maturity determi-
nation in sweet corn. - HortScience 14: 546-548, 1979.

*6955 - DREWITT, E.G., MUSCROFT-TAYLOR, K.E.: The effect of sowing date on spring
wheat and barley. - N.Zeal. J. exp. Agr. 6 (1): 31-33, 1978.

*6956 - DREWS, M.: Der Einfluss einiger Wachstumsfaktoren auf der Wasserhaushalt der
Gewächshausgurke. - Arch. Gartenbau 27: 399-410, 1979.

*6957 - DREWS, M., HOLZ, I.: Messmethodische Untersuchungen zur Bestimmung des Wasser-
zustandes der Gewächshausgurke mit der β-Strahlenabsorptionsmethode. - Arch.
Gartenbau 27: 315-323, 1979.

6958 - DROMGOOLE, F.I.: Desiccation resistance of intertidal and subtidal algae. -
Bot. mar. 23: 149-159, 1980.

6959 - DÜRINIG, H.: Stomatafrequenz bei Blättern von *Vitis*-Arten und -Sorten. -
Vitis 19: 91-98, 1980.

*6960 - DUTHION, C.: Pouvoir réducteur des racines des plantes cultivées. Ses modi-
fications par une courte période d'excès d'eau. - Ann. agron. 30: 323-327,
1979.

*6961 - DUYSEN, M.E., FREEMAN, T.P.: The photosystem and ultrastructural modifications in wheat chloroplasts that developed under slight water stress. - Proc. North Dakota Acad. Sci. 30: 16, 1976.

*6962 - DYER, J.A., BAIER, W.: An index for soil moisture drying patterns. - Can. agr. Eng. 21: 117-118, 1979.

6963 - DYLLA, A.S., TIMMONS, D.R., SHULL, H.: Estimating water used by irrigated corn in west central Minnesota. - Soil Sci. Soc. Amer. J. 44: 823-827, 1980.

6964 - EASTIN, E.F.: Preharvest desiccants for rice. - Crop Sci. 20: 389-391, 1980.

6965 - ECKERT, R.T., HOUSTON, D.B.: Photosynthesis and needle elongation response of *Pinus strobus* clones to low level sulfur dioxide exposures. - Can. J. Forest Res. 10: 357-361, 1980.

6966 - ECKL, K., GRULER, H.: Phase transitions in plant cuticles. - Planta 150: 102-113, 1980.

6967 - EFIMOV, I.T., NAUMENKO, V.P.: Zavisimost' urozhaĭnosti i kachestva oroshaemoĭ vysokolizinovoĭ kukuruzy ot form azotnykh udobreniĭ. [Dependence of productivity and quality of irrigated highlysine corn on nitrogen fertilization.] - Sel'skokhoz. Biol. 15: 120-122, 1980. [In R, ab: E.]

6968 - EHLERINGER, J.: Leaf morphology and reflectance in relation to water and temperature stress. - In: TURNER, N.C., KRAMER, P.J. (ed.): Adaptation of Plants to Water and High Temperature Stress. Pp. 295-308. John Wiley & Sons, New York - Chichester - Brisbane - Toronto 1980.

*6969 - EHLIG, C.F., LeMERT, R.D.: Water use and yields of sugarbeets over a range from excessive to limited irrigation. - Soil Sci. Soc. Amer. J. 43: 403-407, 1979.

6970 - EICKMEIER, W.G.: Photosynthetic recovery of resurrection spikemosses from different hydration regimes. - Oecologia 46: 380-385, 1980.

6971 - EL-BANOBY, F.E., RUDOLPH, K.: Purification of extracellular polysaccharides from *Pseudomonas phaseolicola* which induce water-soaking in bean leaves. - Physiol. Plant Pathol. 16: 425-437, 1980.

*6972 - ELEUTERIUS, L.N., ELEUTERIUS, C.K.: Tide levels and salt marsh zonation. - Bull. Mar. Sci. 29: 394-400, 1979.

*6973 - EL-FORGANY, M., MAKUS, D.J.: Effect of water stress on seed yield and quality of the sweet corn inbred 'Luther Hill'. - J. Amer. Soc. hort. Sci. 104: 102-104, 1979.

6974 - ELIÁŠ, P.: Transpiračné odpory listov a ich meranie. [Transpiration resistances of leaves and their measurements.] - Acta ecol. (Bratislava) 7 (19): 53-115, 1978/1980. [In Slov, ab: R,E.]

6975 - ELIÁŠ, P.: Gradients of several leaf characteristics on stems of two forest herbs. - Biol. Plant. 22: 42-49, 1980.

6976 - ELIÁŠ, P.: Maximálna vodivosť prieduchov rastlín dubovo-hrabového lesa. [Maximum stomata conductance in plants of an oak-hornbeam forest.] - In: Dny Rostlinné Fyziologie II. Pp. 158-161. Vysoká Škola Zemědělská, Brno 1980. [In Slov, ab: E, R.]

6977 - ELIÁŠ, P.: Určovanie odporov vrstiev lesného porastu. [Leaf resistance of canopy layers in a forest stand.] - Folia dendrol. 7: 121-148, 1980. [In Slov, ab: R,G,E.]

*6978 - ELKIEY, T., ORMROD, D.P.: Leaf diffusion resistance responses of three petunia
         cultivars to ozone and/or sulfur dioxide. - Air Pollut. Cont. Assoc. (APCA) J.
         29: 622-625, 1979.

6979 - ELKIEY, T., ORMROD, D.P.: Sorption of ozone and sulfur dioxide by petunia
       leaves. - J. Environ. Qual. 9: 93-95, 1980.

*6980 - ELLIOTT, D.C.: Analysis of variability in the Amaranthus bioassay for cyto-
         kinins. Effects of water stress on benzyladenine- and fusicoccin-dependent
         responses. - Plant Physiol. 63: 269-273, 1979.

6981 - ELLIS, R.H., ROBERTS, E.H.: The influence of temperature and moisture on seed
       viability period in barley (Hordeum distichum L.). - Ann. Bot. 45: 31-37,
       1980.

*6982 - EMERSON, W.W.: Physical properties and structure. - In: RUSSELL, J.S.,
         GREACEN, E.L. (ed.): Soil Factors in Crop Production in a Semi-Arid Environ-
         ment. Pp. 78-104. University of Queensland Press, St. Lucia 1977.

6983 - EPHRITIKHINE, G., LAMANT, A., HELLER, R.: Influence of sucrose on the cha-
       racterization of higher plant membranes by β-glucan synthetase activity and
       its relation to osmotic pressure. - Plant Sci. Lett. 19: 55-64, 1980.

6984 - EPSTEIN, E.: Responses of plants to saline environments. - In: RAINS, D.W.,
       VALENTINE, R.C., HOLLAENDER, A. (ed.): Genetic Engineering of Osmoregulation:
       Impact on Plant Productivity for Food, Chemicals, and Energy. Pp. 7-21.
       Plenum Press, New York - London 1980.

*6985 - ERDEI, L., KUIPER, P.J.C.: The effect of salinity on growth, cation content,
         Na⁺-uptake and translocation in salt-sensitive and salt-tolerant Plantago
         species. - Physiol. Plant. 47: 95-99, 1979.

6986 - ERDEI, L., STUIVER, B.(C.E.E.), KUIPER, P.J.C.: The effect of salinity on
       lipid composition and on activity of Ca^{2+}- and Mg^{2+}-stimulated ATPases in
       salt-sensitive and salt-tolerant Plantago species. - Physiol. Plant. 49:
       315-319, 1980.

*6987 - ERICKSON, P.I., KIRKHAM, M.B., ADJEI, G.B.: Water relations, growth and
         yield of tall and short wheat cultivars irradiated with X-rays. - Environ.
         exp. Bot. 19: 349-356, 1979.

6988 - ERKAN, Z., BANGERTH, F.: Untersuchungen über den Einfluss von Phytohormonen
       und Wachstumsregulatoren auf den Wasserverbrauch, das Stomataverhalten und
       die Photosynthese von Paprika- und Tomatepflanzen. - Angew. Bot. 54: 207-220,
       1980.

6989 - ERNST, W., LUGTENBORG, T.F.: Vergleichende Ökophysiologie von Juncus arti-
       culatus und Holcus lanatus. - Flora 169: 121-134, 1980.

6990 - EVEN-CHEN, Z., SACHS, R.M.: Photosynthesis as a function of short day induc-
       tion and gibberellic acid treatment in Bougainvillea "San Diego Red". -
       Plant Physiol. 65: 65-68, 1980.

6991 - EVERT, R.F.: Vascular anatomy of angiospermous leaves, with special conside-
       ration of the maize leaf. - Ber. Deut. bot. Ges. 93: 43-55, 1980.

*6992 - FADEEVA, L.G.: Vliyanie sukhoveya na razvivayushchiĭsya kolos vlagoobespe-
         chennoĭ pshenitsy. [Dry wind effect on the developing spike of wheat provided
         with sufficient water.] - Izv. Sib. Otd. Akad. Nauk SSSR, Ser. Biol. Nauk
         1979 (1): 108-115, 1979. [In R, ab: E.]

*6993 - FAIZY, S.E-D.A.: Effect of NPK fertilizers on the kinetics of nutrient influx, cation composition, stem rot infection, and drought resistance of tomato plants. - In: GOODIN, J.R., NORTHINGTON, D.K. (ed.): Arid Land Plant Resources. Pp. 550 -563. Texas Technical University, Lubbock 1979.

6994 - FALKE, H., HAMANN, H.-J.: Ein Beitrag zum Einfluss der Beregnung auf die Mikronährstoffgehalte verschiedener Getreidearten. - Arch. Acker- Pflanzenbau Bodenk. 24: 175-180, 1980.

*6995 - FALKOWSKI, M., KOZŁOWSKI, S., WITKOWSKA, B.: Charakterystyka anatomiczno- -morfologiczna blaszek liściowych odmian *Lolium multiflorum*. [Anatomical and morphological characteristics of *Lolium multiflorum* leaf blades.] - Acta agrobot. 31 (1/2): 85-94, 1978. [In Pol, ab: E.]

6996 - FANJUL, L., JONES, H.G., TREHARNE, K.J.: A portable system for simultaneous measurement of transpiration and CO_2 exchange. - Photosynthesis Res. 1: 83-92, 1980.

6997 - FARDJAH, M., LEMÉE, G., PONTAILLER, J.Y.: Dynamique comparée de l'eau sous hêtraie et dans des coupes nues ou a *Calamagrostis epigeios* en forêt de Fontainebleau. - Bull. Ecol. 11: 11-31, 1980.

6998 - FARQUHAR, G.D., SCHULZE, E.-D., KÜPPERS, M.: Responses to humidity by stomata of *Nicotiana glauca* L. and *Corylus avellana* L. are consistent with the optimization of carbon dioxide uptake with respect to water loss. - Aust. J. Plant Physiol. 7: 315-327, 1980.

*6999 - FARRAR, J.F., RELTON, J., RUTTER, A.J.: Sulphur dioxide and the growth of *Pinus sylvestris*. - J. appl. Ecol. 14: 861-875, 1977.

7000 - FAWUSI, M.O.A., AGBOOLA, A.A.: Soil moisture requirements for germination of sorghum, millet, tomato, and *Celosia*. - Agron. J. 72: 353-357, 1980.

*7001 - FAZYLOVA, S.: Vliyanie faktora vlazhnosti na fotosinteticheskuyu sposobnost' nekotorykh pustynnykh vidov kustarnikov i polukustarnikov. [Effect of humidity factor on photosynthetic activity of some desert shrub species.] - In: Fiziologiya i Biokhimiya Dikorastushchikh Kormovykh Rasteniĭ Uzbekistana. Pp. 119-126. Fan, Tashkent 1975. [In R.]

7002 - FEDERER, C.A.: Paper birch and white oak saplings differ in responses to drought. - Forest Sci. 26: 313-324, 1980.

*7003 - FELLOWS, R.J., PATTERSON, R.P., GROSS, H.D., HARRIS, D.: Recovery from water stress in soybeans: Interaction of net photosynthesis, N-fixation, and dry matter partitioning. - Plant Physiol. 63 (Suppl.): 139, 1979.

7004 - FENNER, M.: Some measurements on the water relations of Baobab trees. - Biotropica 12: 205-209, 1980.

7005 - FENSOM,D.S., SILVERBERG, A.: Changes in electroosmotic measurements in *Nitella translucens* when turgor pressure of cells is reduced by solutions of sucrose and mannitol. - Can. J. Bot. 58: 1418-1420, 1980.

*7006 - FERERES, E., CRUZ-ROMERO, G., HOFFMAN, G.J., RAWLINS, S.L.: Recovery of orange trees following severe water stress. - J. appl. Ecol. 16: 833-842, 1979.

7007 - FERMOR, T.R., EGGINS, H.O.W.: The effect of water activity on growth of *Streptomyces* species. - Int. Biodeterioration Bull. 16: 95-101, 1980.

7008 - FERRAR, P.J.: Environmental control of gas exchange in some savanna woody species. I. Controlled environment studies of *Terminalia sericea* and *Grewia flavescens*. - Oecologia 47: 204-212, 1980.

7009 - FERRARIS, R., SINCLAIR, D.F.: Factors affecting the growth of *Pennisetum pur-pureum* in the wet tropics. I. Short-term growth and regrowth. - Aust. J. agr. Res. 31: 899-913, 1980.

7010 - FETCHER, N., TRLICA, M.J.: Influence of climate on annual production of seven cold desert forage species. - J. Range Manage. 33: 35-37, 1980.

7011 - FEYEN, J., BELMANS, C., HILLEL, D.: Comparison between measured and simulated plant water potential during soil water extraction by potted ryegrass. - Soil Sci. 129: 180-185, 1980.

7012 - FINK, D.H., FRASIER, G.W., COOLEY, K.R.: Water harvesting by wax-treated soil surfaces: Progress, problems, and potential. - Agr. Water Manage. 3: 125-134, 1980.

7013 - FINN, G.A., BRUN, W.A.: Water stress effects on CO_2 assimilation, photosyntha-te partitioning, stomatal resistance, and nodule activity in soybean. - Crop Sci. 20: 431-434, 1980.

7014 - FISCHER, R.A.: Influence of water stress on crop yield in semiarid regions. - In: TURNER, N.C., KRAMER, P.J. (ed.): Adaptation of Plants to Water and High Temperature Stress. Pp. 323-339. John Wiley & Sons, New York - Chichester - Brisbane - Toronto 1980.

*7015 - FISCHER, R.A., WOOD, J.T.: Drought resistance in spring wheat cultivars. III Yield associations with morpho-physiological traits. - Aust. J. agr. Res. 30: 1001-1020, 1979.

7016 - FISHER, M.J.: The influence of water stress on nitrogen and phosphorus uptake and concentrations in Townsville stylo (*Stylosanthes humilis*). - Aust. J. exp. Agr. anim. Husb. 20: 175-180, 1980.

*7017 - FLAIG, W.: Modelluntersuchungen zur Beeinflussung des Wasserhaushaltes durch Bestandteile der organischen Bodensubstanz. - Agrochimica 19: 160-163, 1975

7018 - FORD, E.D.: Impact of environment on the shoot elongation of conifers: short term effects. - In: LITTLE, C.H.A. (ed.): Control of Shoot Growth in Trees. Pp. 107-125. Maritimes Forest Research Center, Fredericton 1980.

*7019 - FÖRSTEL, H.: The enrichment of ^{18}O in leaf water under natural conditions. - Radiat. Environ. Biophys. 15: 323-344, 1978.

*7020 - FÖRSTEL, H., HÜTZEN, H.: Variation des $^{18}O/^{16}O$-Verhältnisses im Wasser von Astproben. - Ber.Kernforschungsanlage Jülich Nr. 1595: 1-41, 1979.

*7021 - FÖRSTEL, H., PRAST, H.: Abhängigkeit des $^{18}O/^{16}O$ Verhältnisses im Blattwasser von den physikalischen Bedingungen der Umgebung. 1. Versuch in der Klima-kammer. - Ber. Kernforschungsanlage Jülich Nr. 1619: 1-34, 1979.

7022 - FOSTER, A.C., MAUN, M.A.: Effect of two relative humidities on foliar absorp-tion of NaCl. - Can. J. Plant Sci. 60: 763-766, 1980.

*7023 - FOWLER, D.B., CARLES, R.J.: Growth, development, and cold tolerance of fall--acclimated cereal grains. - Crop Sci. 19: 915-922, 1979.

*7024 - FOWLER, J.L.: Laboratory and field response of preconditioned upland cotton-seed to minimal germination temperatures. - Agron. J. 71: 223-228, 1979.

7025 - FRANCOIS, L.E., CLARK, R.A.: Salinity effects on yield and fruit quality of 'Valencia' orange. - J. Amer. Soc. hort. Sci. 105: 199-202, 1980.

*7026 - FRANCOIS, L.E., MAAS, E.V.: Plant Responses to Salinity: An Indexed Biblio-graphy. - U.S. Department of Agriculture, Berkeley 1978.

7027 - FRANK, A.B.: Photosynthesis, transpiration, and ribulose bisphosphate carbo-xylase of selected crested wheatgrass plants. - Agron. J. 72: 313-316, 1980.

*7028 - FREEMAN, B., ALBRIGO, L.G., BIGGS, R.H.: Ultrastructure and chemistry of cuticular waxes of developing *Citrus* leaves and fruits. - J. Amer. Soc. hort. Sci. 104: 801-808, 1979.

7029 - FREEMAN, B.M., LEE, T.H., TURKINGTON, C.R.: Interaction of irrigation and pruning level on grape and wine quality of Shiraz vines. - Amer. J. Enol. Viticult. 31: 124-135, 1980.

7030 - FRITZ, E.: Microautoradiographic localization of assimilates in phloem: problems and new methods. - Ber. Deut. bot. Ges. 93: 109-121, 1980.

7031 - FROMMHOLD, I.: Struktur und Function der Gramineen-Stomata. - In: HOFFMANN, P., HIEKE, B. (ed.): Biophysik, Biochemie und Physiologie der Photosynthese. Pp. 165-180. Humboldt-Universität, Berlin 1980.

*7032 - FUCIK, J.E., NORWINE, J.: Climatological parameters and grapefruit size rela-tionships in the Rio Grande Valley of Texas. - J. Rio Grande Valley hort. Sci. 33: 83-89, 1979.

7033 - FUHRER, J., ERISMANN, K.H.: Uptake of NO_2 by plants grown at different salini-ty levels. - Experientia 36: 409-410, 1980.

7034 - FUHRER, J., ERISMANN, K.H.: Tolerance of *Aesculus hippocastanum* L. to foliar accumulation of chloride affected by air pollution. - Environ. Pollut. 21: 249-254, 1980.

7035 - GAFF, D.F.: Protoplasmic tolerance of extreme water stress. - In: TURNER, N.C., KRAMER, P.J. (ed.): Adaptation of Plants to Water and High Temperature Stress. Pp. 207-230. John Wiley & Sons, New York - Chichester - Brisbane - Toronto 1980.

*7036 - GALATIS, B., MITRAKOS, K.: On the differential divisions and preprophase microtubule bands involved in the development of stomata of *Vigna sinensis*. - J. Cell Sci. 37: 11-37, 1979.

7037 - GALATIS, B., MITRAKOS, K.: The ultrastructural cytology of the differentiating guard cells of *Vigna sinensis*. - Amer. J. Bot. 67: 1243-1261, 1980.

*7038 - GALE, J., BOLL, W.G.: Growth of bean cells in suspension culture in the pre-sence of NaCl and protein-stabilizing factors. - Can. J. Bot. 57: 777-782, 1979.

*7039 - GALE, J., EASTON, J.: The effect of limestone dust on vegetation in an area with a Mediterranean climate. - Environ. Pollut. 19: 89-102, 1979.

*7040 - GALLAGHER, J.L.: Growth and element compositional responses of *Sporobolus virginicus* (L.) Kunth to substrate salinity and nitrogen. - Amer. Midl. Natur. 102: 68-75, 1979.

*7041 - GAMZIKOVA, O.I., GUDINOVA, L.G.: Vliyanie zasukhi na pokazateli adsorbiruyush-cheĭ poverkhosti kornevoĭ sistemy pshenitsy. [Effect of drought on parameters of wheat root system absorbing surface.] - In: Vsesoyuznoe Soveshchanie Fiziologo-Biokhimicheskie I Ékologicheskie Aspekty Ustoĭchivosti Rasteniĭ k Neblagopriyatnym Faktoram Vneshneĭ Sredy. Pp. 32-33. Sibirskiĭ Institut Fiziologii I Biokhimii Rasteniĭ SO AN SSSR, Irkutsk 1976. [In R.]

7042 - GARLAND, J.A.: The absorption and evaporation of tritiated water vapor by soil and grassland. - Water Air Soil Pollut. 13: 317-333, 1980.

7043 - GARWOOD, E.A., SALETTE, J., LEMAIRE, G.: The influence of water supply to grass on the response to fertilizer nitrogen and nitrogen recovery. - In: The Role of Nitrogen in Intensive Grassland Production. Pp. 59-65. Pudoc, Wageningen 1980.

7044 - GASKELL, M.L., PEARCE, R.B.: Photosynthetic acclimation of maize to solar radiation level. - Maydica 25: 55-64, 1980.

*7045 - GAUCH, H.G.,Jr., STONE, E.L.: Vegetation and soil pattern in a mesophytic forest at Ithaca, New York. - Amer. Midl. Natur. 102: 332-345, 1979.

*7046 - GAUGER, B., BENTRUP, F.W.: A study of dielectric membrane breakdown in the *Fucus* egg. - J. Membrane Biol. 48: 249-264, 1979.

*7047 - GAY, L.W., SAMMIS, T.W.: Estimating phreatophyte transpiration. - Hydrol. Water Resour. Arizona Southwest 7: 133-139, 1977.

7048 - GAZEAU, C., DEREUDDRE, J.: Effets de refroidissements prolongés sur la teneur en eau et la résistance au froid des tissues de la graine de Troène (*Ligustrum vulgare* L. Oléacées). - C.R.Acad. Sci. Paris, Sér. D 290: 1443-1446, 1980.

*7049 - GEDENIDZE, A.A.: Meliorativnaya rol' ol'khovykh lesov v niznennoĭ chasti Kolkhidy. [The land-reclamation role of *Alnus barbata* forest in low-lying part of Kolkhida.] - In: Voprosy Gornogo Lesovedeniya v Gruzii. Vol. 25. Pp. 52-55. Sabchota Adzhara, Batumi 1976. [In R, ab:E.]

7050 - GEIGER, D.R., FONDY, B.R.: Response of phloem loading and export to rapid changes in sink demand. - Ber. Deut. bot. Ges. 93: 177-186, 1980.

*7051 - GENKEL', P.A.: O sopryazhennoĭ i konvergentnoĭ ustoĭchivosti rasteniĭ. [Coupled and convergent resistance of plants to unfavourable environment.] - Fiziol. Rast. 26: 921-931, 1979. [In R, ab: E.]

7052 - GENKEL', P.A., SHELAMOVA, N.A.: Diagnostika zharo- i zasukhoustoĭchivosti rasteniĭ metodom fotometricheskogo opredeleniya statolitnogo krakhmala. [Heat and drought resistance diagnosis using method of statolite starch photometric determination.] - Sel'skokhoz. Biol. 15: 448-451, 1980. [In R, ab: E.]

7053 - GEORGIEV, T., MUKHTANOV, I.: Zavisimosti mezhdu s"d"rzhanieto na vlaga v z"rnoto pri pribiraneto i nyakoi priznatsi na tsarevitsata. [Correlations between kernel moisture content at harvest time and some maize characters.] - Genet. Selek. (Sofia) 13: 180-190, 1980. [In Bulg, ab: E,R.]

*7054 - GEORGIEVA, V., KUDREV, T.: Changes in nitratereductase activity in maize plants under conditions of osmotic stress and the influence of pre-seeding boric acid tratment. - In: KUDREV, T., STOYANOV, I., GEORGIEVA, V. (ed.): Mineral Nutrition of Plants. Vol. II. Pp. 231-235. Publishing House, Central Cooperative Union, Sofia 1979.

7055 - GEPSTEIN, S., THIMANN, K.V.: Changes in the abscisic acid content of oat leaves during senescence. - Proc. nat. Acad. Sci. USA 77: 2050-2053, 1980.

7056 - GERAKIS, P.A., PAPAKOSTA-TASOPOULOU, D.: Effects of dense planting and artificial shading on five maize hybrids. - Agr. Meteorol. 21: 129-137, 1980.

7057 - GERDENITSCH, W.: Der Einfluss von Isoptin (Verapamil) auf die Wasserpermeabilität der Innenepidermiszellen von *Allium cepa*. - Phyton 20: 235-249, 1980.

*7058 - GESALMAN, C.M., DAVIS, D.D.: Ozone susceptibility of ten azalea cultivars as related to stomatal frequency or conductance. - J. Amer. Soc. hort. Sci. 103: 489-491, 1978.

*7059 - GHERMAN, N.: Regimul de irigare la semincerii de pătrunjel. [The irrigation
        system for the parsley seed plants.] - An. Inst. Cercetări Pentru Legumino-
        cult. Floricult. 5: 47-50, 1979. [In Roum, ab: E,F.]

*7060 - GHOSE, M.: Ontogenetic study of stomata and trichomes in some palms. - Phyto-
        morphology 29: 26-33, 1979.

 7061 - GHOSH, R.K.: Estimation of soil-moisture characteristics from mechanical pro-
        perties of soils. - Soil Sci. 130: 60-61, 1980.

*7062 - GICHNER, T., VELEMÍNSKÝ, J.: Post-treatment modulation affecting the yield of
        chromosomal aberrations induced by diethyl sulphate and methyl methanesulpho-
        nate in barley. - Mutat. Res. 60: 181-187, 1979.

 7063 - GILLEY, J.R., MARTIN, D.L., SPLINTER, W.E.: Application of a simulation model
        of corn growth to irrigation management decisions. - In: YARON, D., TAPIERO,
        C. (ed.): Operations Research in Agriculture and Water Resources. Pp. 485-
        500. North-Holland Publishing Company, IFORS, Amsterdam 1980.

 7064 - GISI, U., ZENTMYER, G.A., KLURE, L.J.: Production of sporangia by *Phytophtho-
        ra cinnamomi* and *P. palmivora* in soils at different matric potentials. - Phy-
        topathology 70: 301-306, 1980.

*7065 - GOLOMAZOVA, G.M.: Optimal'nye usloviya fotosinteza listvennitsy sibirskoĭ.
        [Optimum conditions for photosynthesis of *Larix sibirica*.] - In: Fiziologo-
        -Biokhimicheskie Protsessy u Khvoĭnykh Rasteniĭ. Pp. 24-34, 142. Akad. Nauk
        SSSR, Krasnoyarsk 1978. [In R.]

 7066 - GOMES, A.R.S., KOZLOWSKI, T.T.: Growth responses and adaptations of *Fraxinus
        pennsylvanica* seedlings to flooding. - Plant Physiol. 66: 267-271, 1980.

*7067 - GÓMEZ-LEPE, B.E., LEE-STADELMANN, O., PALTA, J.P., STADELMANN, E.J.: Effects
        of octylguanidine on cell permeability and other protoplasmic properties of
        Allium cepa epidermal cells. - Plant Physiol. 64: 131-138, 1979.

*7068 - GOMM, F.B.: Herbage yield and nitrate concentration in meadow plants as
        affected by environmental variables. - J. Range Manage. 32: 359-364, 1979.

 7069 - GOMM, F.B.: Correlation of environmental factors with nitrate concentration
        in meadow plants. - J. Range Manage. 33: 223-228, 1980.

 7070 - GOOD, N.E., BELL, D.H.: Photosynthesis, plant productivity, and crop yield. -
        In: CARLSON, P.S. (ed.): The Biology of Crop Productivity. Pp. 3-50. Academic
        Press, New York - London - Toronto - Sydney - San Francisco 1980.

 7071 - GOODWIN, T.W.: The Biochemistry of the Carotenoids. Vol. I. Plants. 2nd Ed. -
        Chapman and Hall, London - New York 1980.

*7072 - GORDON, I.L., BALAAM, L.N., DERERA, N.F.: Selection against sprouting damage
        in wheat. II. Harvest ripeness, grain maturity and germinability. - Aust. J.
        agr. Res. 28: 583-599, 1977.

*7073 - GOTTSCHALK, K.W., DICKMANN, D.I.: Environmental effects on photosynthesis
        and stomatal conductance of four, two-year-old *Populus* clones grown in the
        field. - Plant Physiol. 63 (Suppl.): 127, 1979.

*7074 - GOUDRIAAN, J., Van KEULEN, H.: The direct and indirect effects of nitrogen
        shortage on photosynthesis and transpiration in maize and sunflower. - Neth.
        J. agr. Sci. 27: 227-234, 1979.

 7075 - GOUNOT, M., YU, O.: Recherches sur l'évaluation de la productivité primaire
        épigée des graminées prairiales. - Acta oecol. - Oecol. Plant. 1: 81-102,
        1980.

7076 - GOVINDJEE, DOWNTON, W.J.S., FORK, D.C., ARMOND, P.A.: Chlorophyll *a* fluorescence transient as an indicator of water potential of leaves. - Carnegie Inst. Washington Year Book 79: 191-193, 1980.

7077 - GOWIN, T., LOURTIOUX, A., MOUSSEAU, M.: Influence of constant growth temperature upon the productivity and gas exchange of seedlings of Scots pine and European larch. - Forest Sci. 26: 301-309, 1980.

7078 - GRACE, J., FASEHUN, F.E., DIXON, M.: Boundary layer conductance of the leaves of some tropical timber trees. - Plant Cell Environ. 3: 443-450, 1980.

*7079 - GRACE, J., MARKS, T.C.: Physiological aspects of bog production at Moor House. - In: HEAL, O.W., PERKINS, D.F. (ed.): Production Ecology of British Moors and Montane Grasslands. Pp. 38-51. Springer-Verlag, Berlin - Heidelberg - New York 1978.

*7080 - GRAFE, V.B.: Schätzung der potentiellen Evaporation aus meteorologischen Grössen. - Arch. Acker- Pflanzenbau Bodenk. 23: 701-705, 1979.

7081 - GRAHAM, C.W.: The effects of rainfall and soil type on the population dynamics of cereal cyst nematode (*Heterodera avenae*) on spring barley (*Hordeum vulgare*) and spring oats (*Avena sativa*). - Ann. appl. Biol. 94: 243-253, 1980.

7082 - GRANGE, A.: Vieillissement des graines de *Phaseolus vulgaris* (L.) var. Contender I. Effets sur la germination, la vigueur, la teneur en eau et la variation des formes d'azote. - Physiol. vég. 18: 579-586, 1980.

*7083 - GREB, B.W.: Reducing drought effects on croplands in the west-central Great Plains. - Agr. Inform. Bull. 420: 1-28, 1979.

*7084 - GREB, B.W., SMIKA, D.E., WELSH, J.R.: Technology and wheat yields in the Central Great Plains. Experiment station advances. - J. Soil Water Conserv. 34: 264-268, 1979.

7085 - GREENE, D.M., KIRKHAM, M.B.: Water-conserving wheat irrigation schedules based on climatic records. - Irrig. Sci. 1: 241-246, 1980.

*7086 - GREENE, D.M., SUTHERLAND, S.M., KIRKHAM, M.B.: Influence of area on winter wheat climatic models. - Climatic Change 2: 21-32, 1979.

7087 - GREENWAY, H., MUNNS, R.: Mechanisms of salt tolerance in nonhalophytes. - Annu. Rev. Plant Physiol. 31: 149-190, 1980.

*7088 - GREENWOOD, E.A.N., BERESFORD, J.D.: Evaporation from vegetation in landscapes developing secondary salinity using the ventilated-chamber technique. I. Comparative transpiration from juvenile *Eucalyptus* above saline ground-water seeps. - J. Hydrol. 42: 369-382, 1979.

7089 - GREENWOOD, E.A.N., BERESFORD, J.D.: Evaporation from vegetation in landscapes developing secondary salinity using the ventilated-chamber technique. II. Evaporation from *Atriplex* plantations over a shallow saline water table. - J. Hydrol. 45: 313-319, 1980.

7090 - GRIFFIN, R.E., STUTLER, R.K.: Pump power. Energy savings in efficient irrigation system management. - Utah Sci. 41: 64-65, 1980.

7091 - GROOTJANS, A.P., TEN KLOOSTER, W.P.: Changes of ground water regime in wet meadows. - Acta bot. Neerl. 29: 541-554, 1980.

7092 - GROSS, K.: Öffnungszustand der Stomata, Xylem-Wasserpotential und Netto-Photosynthese junger Fichten vor und nach der Verpflanzung. - Forstwis. Centralbl. 99: 12-21, 1980.

7093 - GROSS, K., PHAM-NGUYEN, T., UNGER, H.: Tägliche und saisonale Änderungen des Wasserpotentials und seiner Komponenten in den Kronen von Fichten unterschiedlichen Alters. - Allg. Forst. Jagdzeit. 151(4/5): 69-80, 1980.

*7094 - GRZESIAK, S.: Influence of sulphur dioxide on the relative rate of photosynthesis in four species of cultivated plants under optimum soil moisture and drought conditions. - Bull. Acad. Pol. Sci., Ser. Sci. biol. 27: 309-321, 1979.

*7095 - GUILLOT, F.S., DROSTE, T., HART, W.: Egg attachment of citrus blackfly *Aleurocanthus woglumi* Ashby to citrus leaves. - Southwest Entomol. 4: 167-169, 1979.

7096 - GULATI, K.L., OSWAL, M.C., NAGPAUL, K.K.: Effect of concentration of boron on the uptake and yield of tomato and wheat at different levels of irrigation. - Plant Soil 54: 479-484, 1980.

7097 - GUNTON, J.L., EVENSON, J.P.: Moisture stress in navy beans I. Effect of withholding irrigation at different phenological stages on growth and yield. - Irrig. Sci. 2: 49-58, 1980.

7098 - GUNTON, J.L., EVENSON, J.P.: Moisture stress in navy beans II. Relationship between leaf water potential and growth and yield. - Irrig. Sci. 2: 59-65, 1980.

*7099 - GUPTA, J.P., AGGARWAL, R.K., KAUL, P.: Effect of the application of pond sediments on soil properties and yield of pearlmillet and greengram in arid areas of western Rajasthan. - Ind. J. agr. Sci. 49: 875-879, 1979.

7100 - GUPTA, R.K.: Intraspecific difference in the leakage of photosynthates in *Acrocladium cuspidatum* (Hedw.) Lindb. - Ind. J. exp. Biol. 18: 1055-1057, 1980.

*7101 - GUSTA, L.V., FOWLER, D.B.: Cold resistance and injury in winter cereals. - In: MUSSELL, H., STAPLES, R. (ed.): Stress Physiology in Crop Plants. Pp. 160 -178. John Wiley & Sons, New York 1979.

*7102 - HAHN, S.K.: Sweet potato. - In: ALVIM, P.de T., KOZLOWSKI, T.T. (ed.): Ecophysiology of Tropical Crops. Pp. 237-248. Academic Press, New York - San Francisco - London 1977.

*7103 - HALE, M.G., MOORE, L.D.: Factors affecting root exudation II: 1970-1978. - Adv. Agron. 31: 93-124, 1979.

7104 - HALL, A.E., SCHULZE, E.-D.: Stomatal response to environment and a possible interrelation between stomatal effects on transpiration and CO_2 assimilation. - Plant Cell Environ. 3: 467-474, 1980.

7105 - HALL, A.E., SCHULZE, E.-D.: Drought effects on transpiration and leaf water status of cowpea in controlled environments. - Aust. J. Plant Physiol. 7: 141-147, 1980.

7106 - HALL, A.J., GINZO, H.D., LEMCOFF, J.H., SORIANO, A.: Influence of drought during pollen-shedding on flowering, growth and yield of maize. - Z. Acker-Pflanzenbau. 149: 287-298, 1980.

7107 - HALLAM, N.D., LUFF, S.E.: Fine structural changes in the mesophyll tissue of the leaves of *Xerophyta villosa* during desiccation. - Bot. Gaz. 141: 173-179, 1980.

7108 - HALLAM, N.D., LUFF, S.E.: Fine structural changes in the leaves of the desiccation-tolerant plant *Talboltia elegans* during extreme water stress. - Bot. Gaz. 141: 180-187, 1980.

7109 - HALVORSON, A.D., REULE, C.A.: Alfalfa for hydrologic control of saline seeps.
- Soil Sci. Soc. Amer. J. 44: 370-374, 1980.

7110 - HAMZA, M.: Réponses des végétaux à la salinité. - Physiol. vég. 18: 69-81,
1980.

7111 - HANKS, R.J., ASHCROFT, G.L.: Applied Soil Physics. Soil Water and Temperature
Applications. (Advanced Series in Agricultural Sciences 8). - Springer-Verlag,
Berlin - Heidelberg - New York 1980.

7112 - HANKS, R.J., PUCKRIDGE, D.W.: Prediction of the influence of water, sowing
date and planting density on dry matter production of wheat. - Aust. J. agr.
Res. 31: 1-11, 1980.

*7113 - HANSCOM, Z.,III, JOHNSON, H.B., HOFFMAN, G.J.: The effect of environmental
stress on the salt tolerance and gas exchange of pinto bean. - Plant Physiol.
63 (Suppl.): 60, 1979.

7114 - HANSEN, V.: Estimation of potential evapotranspiration. A theoretical approach.
- Meld. Norges Landbrukshøgsk. 59(10): 1-12, 1980.

7115 - HANSEN, V.: On the problem of estimating daily rates of potential evapotranspiration. - Meld. Norges Landbrukshøgsk. 59(15): 1-15, 1980.

7116 - HANSON, A.D.: Interpreting the metabolic responses of plants to water stress.-
HortScience 15: 623-629, 1980.

7117 - HANSON, A.D., NELSEN, C.E.: Water: Adaptation of crops to drought-prone
environments. - In: CARLSON, N.E. (ed.): The Biology of Crop Productivity.
Pp. 77-152. Academic Press, New York - London - Toronto - Sydney - San
Francisco 1980.

*7118 - HANSON, A.D., NELSEN, C.E., PEDERSEN, A.R., EVERSON, E.H.: Capacity for pro-
line accumulation during water stress in barley and its implications for
breeding for drought resistance. - Crop Sci. 19: 489-493, 1979.

7119 - HANSON, A.D., SCOTT, N.A.: Betaine synthesis from radioactive precursors in
attached,water-stressed barley leaves. - Plant Physiol. 66: 342-348, 1980.

7120 - HANSON, J.D., DYE, A.J.: Diurnal and seasonal patterns of photosynthesis of
honey mesquite. - Photosynthetica 14: 1-7, 1980.

7121 - HARAUX, F., KOUCHKOVSKY, Y.,de: Measurement of chloroplast internal protons
with 9-aminoacridine. Probe binding, dark proton gradient, and salt effects.
- Biochem. biophys. Acta 592: 153-168, 1980.

7122 - HARDY, J.R.: Survey of methods for the determination of soil moisture content
by remote sensing methods. - In: FRAYSSE, G. (ed.): Remote Sensing Application in Agriculture and Hydrology. Pp. 233-247. A.A. Balkema, Rotterdam 1980.

*7123 - HARRIS, J.M., McCONCHIE, D.L., POVEY, W.A.: Wood properties of clonal radiata
pine grown in soils with different levels of available nitrogen, phosphorus
and water. - N.Zeal. J. Forest. Sci. 8: 417-430, 1978.

7124 - HARRISON, J.G.: Effects of environmental factors on growth of lesions on
field bean leaves infected by *Botrytis fabae*. - Ann. appl. Biol. 95: 53-61,
1980.

7125 - HARRISON, J.G.: The production of toxins by *Botrytis fabae* in relation to
growth of lesions on field bean leaves at different humidities. - Ann. appl.
Biol. 95: 63-71, 1980.

*7126 - HART, G.E., LOMAS, D.A.: Effects of clearcutting on soil water depletion in
         an Engelman spruce stand. - Water Resour. Res. 15: 1598-1602, 1979.

7127 - HART, W.E., NORUM, D.I., PERI, G.: Optimal seasonal irrigation application
       analysis. - J. Irrig. Drain. Div. ASCE 106(IR3): 221-235, 1980.

7128 - HARVEY, D.M.: Seed production in leafless and conventional phenotypes of *Pi-
       sum sativum* L. in relation to water availability within a controlled environ-
       ment. - Ann. Bot. 45: 673-680, 1980.

7129 - HASPELOVÁ-HORVATOVIČOVÁ, A., HOLÚBKOVÁ, B.: Obsah chlorofylu v experimentálne
       vädnúcich listoch. [Chlorophyl content in experimentally wilted leaves.] -
       In: Dny Rostlinné Fyziologie II. Pp. 183-186. Vysoká Škola Zemědělská, Brno
       1980. [In Slov, ab: E,R.]

7130 - HAVIS, J.R.: Container moisture state and stomatal resistance in nursery
       plants. - HortScience 15: 638-639, 1980.

7131 - HAVRANEK, W.M.: Das Boden-Wasserpotential - bestimmbar durch Messung des
       Dämmerungs-Wasserpotentials von Jungfichten? - Flora 169: 32-37, 1980.

*7132 - HAYAMA, T., NAKAGAWA, S., TAZAWA, M.: Membrane depolarization induced by
         transcellular osmosis in internodal cells of *Nitella flexilis*. - Protoplasma
         98: 73-90, 1979.

7133 - HAYNES, R., HERRING, S.: Effect of trickle irrigation on yield and quality
       of summer squash. - Arkansas Farm Res. 29(5): 6, 1980.

*7134 - HEARN, A.B.: Crop physiology. - In: ARNOLD, M.H. (ed.): Agricultural Research
         for Development. Pp. 77-122. Cambridge University Press, Cambridge - New York
         1976.

*7135 - HEARN, A.B.: Water relationships in cotton. - Outlook Agr. 10: 159-166,
         1979.

7136 - HEATHERLY, L.G.: Growth of soybeans at different soil matric potentials. -
       Soil Sci. 130: 331-335, 1980.

7137 - HEATHERLY, L.G.: Effect of upper-profile soil water potential on soybean root
       and shoot relationships. - Field Crops Res. 3: 165-171, 1980.

7138 - HEATHERLY, L.G., McMICHAEL, B.L., GINN, L.H.: A weighing lysimeter for use in
       isolated field areas. - Agron. J. 72: 845-847, 1980.

7139 - HEEN, A.: Methods for root studies of annual plants. - Meld. Norges Land-
       brukshøgsk. 59(16): 1-17, 1980.

7140 - HEEN, A.: Root development and water use in some Norwegian barley, wheat and
       oat varieties. - Meld. Norges Landbrukshøgsk. 59 (17): 1-11, 1980.

7141 - HEERMANN, D.F.: Irrigation scheduling. - In: YARON, D., TAPIERO, C. (ed.):
       Operations Research in Agriculture and Water Resources. Pp. 501-516. North-
       -Holland Publishing Company, IFORS, Amsterdam 1980.

*7142 - HEGARTY, T.W.: Seed and soil factors affecting the level and rate of emergen-
         ce. - Acta Hort. 72: 11-25, 1978.

7143 - HEGARTY, T.W., ROSS, H.A.: Investigations of control mechanisms of germina-
       tion under water stress. - Isr. J. Bot. 29: 83-92, 1980.

7144 - HEIKAL, M.M., AHMED, A.M., SHADDAD, M.A.: Changes in dry weight and mineral
       composition of some oil producing plants over a range of salinity stresses. -
       Biol. Plant. 22: 25-33, 1980.

7145 - HEINZE, M., FIEDLER, H.J.: Wasserverbrauch, Ernährung und Wachstum von Kiefern-
sämlingen bei verschiedener Belichtung, Bewässerung und Düngung. - Flora 169:
89-103, 1980.

7146 - HEINZE, M., FIEDLER, H.J.: Zur Reaktion von Lärchensämlingen auf unterschied-
liche Belichtung, Bewässerung und Düngung. - Flora 170: 329-339, 1980.

7148 - HELWEG, O.J., ALVAREZ, D.: Estimating irrigation water quantity and quality. -
J. Irrig. Drain. Div. ASCE 106(IR3): 175-188, 1980.

*7149 - HENCKEL, P.A.: Physiological ways of plant adaptation against drought. -
Agrochimica 19: 431-436, 1975.

7150 - HENNY, R.J.: Relative humidity affects *in vivo* pollen germination and seed
production in *Dieffenbachia maculata* 'Perfection'. - J. Amer. Soc. hort. Sci.
105: 546-548, 1980.

7151 - HEW, C.S., LEE, G.L., WONG, S.C.: Occurrence of non-functional stomata in the
flowers of tropical orchids. - Ann. Bot. 46: 195-201, 1980.

*7152 - HEYDECKER, W., GIBBINS, B.M.: The 'priming' of seeds. - Acta Hort. 83: 213-
223, 1978.

7153 - HICKLENTON, P.R., JOLLIFFE, P.A.: Alterations in the physiology of CO_2 exchange in tomato plants in CO_2-enriched atmospheres. - Can. J. Bot. 58: 2181-2189,
1980.

*7154 - HILL, R.W., JOHNSON, D.R., RYAN, K.H.: A model for predicting soybean yields
from climatic data. - Agron. J. 71: 251-256, 1979.

7155 - HILL, R.W., KELLER, J.: Irrigation system selection for maximum crop profit. -
Trans. ASAE 23: 366-372, 1980.

7156 - HINCKLEY, T.M., DUHME, F., HINCKLEY, A.R., RICHTER, H.: Water relations of
drought hardy shrubs: osmotic potential and stomatal reactivity. - Plant Cell
Environ. 3: 131-140, 1980.

7157 - HIRA, G.S., SINGH, N.T.: Irrigation water requirement for dissolution of
gypsum in sodic soil. - Soil Sci. Soc. Amer. J. 44: 930-933, 1980.

7158 - HO, J., LIN, P., YANG, H.J.: [Studies on the eco-physiological indices of
rubber trees at chilling temperature.] - Acta bot. Sinica 22: 41-48, 1980.
[In Chin, ab: E.]

*7159 - HOCH, H.C.: Penetration of chemicals into the *Malus* leaf cuticle. An ultrastructural analysis. - Planta 147: 186-195, 1979.

*7160 - HOCKING, A.D., PITT, J.I.: Water relations of some *Penicillium* species at
25 ºC. - Trans. Brit. mycol. Soc. 73: 141-145, 1979.

7161 - HOCKING, P.J.: The composition of phloem exudate and xylem sap from tree
tobacco (*Nicotiana glauca* Grah.). - Ann. Bot. 45: 633-643, 1980.

*7162 - HOFFMAN, G.J., RAWLINS, S.L., OSTER, J.D., JOBES, J.A., MERRILL, S.D.:
Leaching requirement for salinity control. I. Wheat, sorghum, and lettuce. -
Agr. Water Manage. 2: 177-192, 1979.

7163 - HOFFMAN, G.J., SHALHEVET, J., MEIRI, A.: Leaf age and salinity influence
water relations of pepper leaves. - Physiol. Plant. 48: 463-469, 1980

7164 - HOFFMANN, P., HIEKE, B. (ed.): Biophysik, Biochemie und Physiologie der Photosynthese.(Colloquia Pflanzenphysiologie Humboldt-Universität Berlin Nr. 3.).
- Humboldt-Universität, Berlin 1980.

7165 - HOLBROOK, F.S., WELSH, J.R.: Soil-water use by semidwarf and tall winter wheat cultivars under dryland field conditions. - Crop Sci. 20: 244-246, 1980.

7166 - HOLDER, C.B., BROWN, K.W.: The relationship between oxygen and water uptake by roots of intact bean plants. - Soil Sci. Soc. Amer. J. 44: 21-25, 1980.

7167 - HOLLINS, T.W., SCOTT, P.R.: Epidemiology of eyespot (*Pseudocercosporella herpotrichoides*) on winter wheat, with particular reference to the period of infection. - Ann. appl. Biol. 95: 19-29, 1980.

*7168 - HORIE, T.: A simulation model for cucumber growth to form basis for managing the plant-environment system. - Acta Hort. 87: 215-224, 1978.

*7169 - HORIE, T.: Simulation of the growth of sunflower plant canopy in relation to solar radiation. - In: MONSI, M., SAEKI, T. (ed.): Ecophysiology of Photosynthetic Productivity. JIBP Synthesis Vol. 19. Pp. 260-267. University of Tokyo Press, 1978.

*7170 - HORIE, T.: Studies on photosynthesis and primary production of rice plants in relation to meteorological environments. II. Gaseous diffusive resistances, photosynthesis and transpiration in the leaves as influenced by atmospheric humidity, and air and soil temperatures. - J. agr. Meteorol. 35: 1-12, 1979.

7171 - HOWARD, B.H.: Moisture change as a component of disbudding responses in studies of supposed relationships between bud activity and rooting in leafless cuttings. - J. hort. Sci. 55: 171-180, 1980.

*7172 - HSIEH, W.H., SNYDER, W.C., SMITH, S.N.: Influence of carbon sources, amino acids, and water potential on growth and sporulation of *Fusarium moniliforme*. - Phytopathology 69: 602-604, 1979.

*7173 - HUANG, P., FÜHRER, E.: Zur Nahrungsqualität von Fichtennadeln für forstliche Schadinsekten. 12. Variabilität der Nadelhautstruktur. - Z. angew. Entomol. 88: 231-245, 1979.

*7174 - HUBER, S.C., MORELAND, D.E.: Permeability properties of the inner membrane of mung bean mitochondria and changes during energization. - Plant Physiol. 64: 115-119, 1979.

*7175 - HUBER, W.: Die Rolle von Abscisinsäure und Cytokininen in Pflanzen unter Stresseinwirkungen. - Ber. Deut. bot. Ges. 92: 193-207, 1979.

*7176 - HUKKERI, S.B., RAMANAIAH, K.V.: Soil and water management problems and practices for rice cultivation in non-traditional areas. - Ind. J. Genet. Plant Breed. 39A: 31-38, 1978.

*7177 - HUKKERI, S.B., SHARMA, A.K.: Tailoring the irrigation schedule for higher water-use efficiency in potato production. - Ind. J. agr. Sci. 49: 336-339, 1979.

7178 - HUKKERI, S.B., SHARMA, A.K.: Irrigation requirement of field-pea for grain. - Ind. J. agr. Sci. 50: 157-160, 1980.

7179 - HUKKERI, S.B., SHARMA, A.K.: Water-use efficiency of transplanted and direct-sown rice under different water-management practices. - Ind. J. agr. Sci. 50: 240-243, 1980.

*7180 - HUNER, N.P.A., PARSONS, L.R., CARTER, J.V.: The structure and activity of RuBP carboxylase from rye exposed to low water potentials. - Plant Physiol. 63 (Suppl.): 139, 1979.

7181 - HUNTER, M.N., JABRUN, P.L.M.,de, BYTH, D.E.: Response of nine soybean lines
to soil moisture conditions close to saturation. - Aust. J. exp. Agr. anim.
Husb. 20: 339-345, 1980.

*7182 - HUNTER, R.B., MORTIMORE, G., GERRISH, E.E., KANNENBERG, L.W.: Field drying
of flint and dent endosperm maize. - Crop Sci. 19: 401-402, 1979.

7183 - HUZULÁK, J.: Water relations of *Crataegus oxyacantha*, *Cornus mas* and *Ligustrum
vulgare*. - Biológia (Bratislava) 35: 39-46, 1980.

7184 - HUZULÁK, J.: Vodný potenciál listov za anticyklonálnych poveternostných situá-
ciī. [Leaves water potential under anticyclonic weather types.] - In: Dny
Rostlinné Fyziologie II. Pp. 202-204. Vysoká Škola Zemědělská, Brno 1980.
[In Slov, ab: E,R.]

7185 - HUZULÁK, J., MATEJKA, F.: Study of xylem pressure potential daily dynamics by
means of autocorrelation analysis. - Biol. Plant. 22: 336-340, 1980.

*7186 - HWANG, Z.-C.: [The effects of drought on the water balance and nitrogen meta-
bolism in sand-fixing plants.] - Acta bot. Sinica 21: 314-319, 1979. [In Chin,
ab: E.]

*7187 - IERUSALIMOV, E.N.: Narushenie fiziologicheskikh protsessov u derev'ev povrezh-
dennykh nasekomymi-defoliatorami. [Disturbance of physiological processes in
trees damaged by insect defoliators.] - Lesovedenie 1979(2): 62-71, 1979. [In
R, ab: E.]

7188 - IGNATOVA, L.K.: Poluchenie protoplastov iz C_3- i C_4-rasteniĭ i ikh fotosinte-
ticheskaya aktivnost'.[Protoplasts from C_3 and C_4 plants and their photosynthetic activity.] - Fiziol. Rast. 27: 80-85, 1980. [In R, ab: E.]

7189 - IL'INS'KA, A.P.: Porivnyal'no-anatomichne doslidzhennya epidermisu lystkiv
deyakykh vydiv rodu *Achillea* L. [Comparative and anatomical studies of leaf
epidermis in some *Achillea* L. species.] - Ukr. bot. Zh. 37(3): 25-29, 1980.
[In Ukr, ab: E.]

*7190 - INABA, T., HINO, T.: Effect of water and temperature on the oogonium and
oospore formation of *Peronospora manshurica* in lesions on *Glycine soja*. -
Ann. phytopathol. Soc. Jap. 45: 468-473, 1979.

7191 - INABA, T., HINO, T.: Influence of water and temperature on oogonium and
oospore formation of downy mildew fungus, *Peronospora manshurica* in soybean
lesions. - Ann. phytopathol. Soc. Jap. 46: 480-486, 1980.

7192 - INOUE, M.T.: Photosynthesis and transpiration in *Cedrela fissilis* Vell seedlings in relation to light intensity and temperature. - Turrialba 30: 280-283,
1980.

7193 - INUYAMA, S.: [Effect of the amount of irrigation water on growth and grain
yield of grain sorghum.] - Jap. J. Crop Sci. 49: 226-231, 1980. [In Jap, ab:
E.]

*7194 - IRUTHAYARAJ, M.R., MORACHAN, Y.B.: Production function analysis of irrigation
water and nitrogen fertilizer in swamp rice production. - Turrialba 27: 411-
413, 1977.

*7195 - IRUTHAYARAJ, M.R., MORACHAN, Y.B.: Effect of season, variety, nitrogen and
water management on water use efficiency in swamp rice. - Ind. J. agr. Res.
12: 47-51, 1978.

7196 - IRUTHAYARAJ, M.R., MORACHAN, Y.B.: Effect of season, water management and
nitrogen on leaf area index and yield of short duration rice varieties. -
Mysore J. agr. Sci. 14: 183-189, 1980.

7197 - IRVINE, R.B., HARVEY, B.L., ROSSNAGEL, B.G.: Rooting capability as it relates to soil moisture extraction and osmotic potential of semidwarf and normal--statured genotypes of six-row barley (*Hordeum vulgare* L.). - Can. J. Plant Sci. 60: 241-248, 1980.

*7198 - ISOBE, S.: Equilibrium evapotranspiration as influenced by the correlation of temperature with its vertical gradient. - J. agr. Meteorol. 35: 85-92, 1979.

7199 - IVANCHENKO, V.M., MARSHAKOVA, M.I., IZBAVITELEV, S.P., KRUCHININA, S.S.: Osobennosti svetoindutsiruemykh izmeneniǐ ob"ema izolirovannykh khloroplastov v svyazi s usloviyami ikh inkubatsii. [Peculiarities of light-induced changes in isolated chloroplast volume in relation to experimental conditions.] - Fiziol. Rast. 27: 696-703, 1980. [In R, ab: E.]

7200 - JACKSON, G.M., VARRIANO-MARSTON, E.: A simple autoradiographic technique for studying diffusion of water into seeds. - Plant Physiol. 65: 1229-1230, 1980.

*7201 - JACKSON, M.B., CAMPBELL, D.J.: Effects of benzyladenine and gibberellic acid on the responses of tomato plants to anaerobic root environments and to ethylene. - New Phytol. 82: 331-340, 1979.

7202 - JACKSON, T.J.: Profile soil moisture from surface measurements. - J. Irrig. Drain. Div. ASCE 106(IR2): 81-92, 1980.

*7203 - JACQUES, G.L., HARVEY, R.G.: Dinitroaniline herbicide phytotoxicity as influenced by soil moisture and herbicide vaporization. - Weed Sci. 27: 536-539, 1979.

7204 - JÄGER, H.-J., KLEIN, H.: Biochemical and physiological effects of SO_2 on plants. - Angew. Bot. 54: 337-348, 1980.

*7205 - JAGGI, I.K., GUPTA, R.K., RUSSELL, M.B.: Seasonal changes in soil-water content and water-potential profiles in fallow land and in fields under irrigated and rainfed wheat. - Ind. J. agr. Sci. 46: 493-502, 1976.

7206 - JAMBHALE, N., NERKAR, Y.S.: A technique for permanent chloroplast preparations. - Curr. Sci. 49: 150, 1980.

7207 - JANNERMANN, G., FUHRMANN, U.: Auswirkung von Erschliessungsanteil und Erzeugnisstruktur auf die Effektivität der Klarwasserberegnung in Pflanzenproduktionsbetrieben. - Arch. Acker- Pflanzenbau Bodenk. 24: 367-374, 1980.

*7208 - JARVIS, M.C., DUNCAN, H.J.: Water loss from potato tuber discs: a method for assessing wound healing. - Potato Res. 22: 69-73, 1979.

7209 - JARVIS, P.G.: Stomatal response to water stress in conifers. - In: TURNER, N.C., KRAMER, P.J. (ed.): Adaptations of Plants to Water and High Temperature Stress. Pp. 105-122. John Wiley & Sons, New York - Chichester - Brisbane - Toronto 1980.

7210 - JARVIS, R.G., MANSFIELD, T.A.: Reduced stomatal responses to light, carbon dioxide and abscisic acid in the presence of sodium ions. - Plant Cell Environ. 3: 279-283, 1980.

7211 - JAYNES, D.B., TYLER, E.J.: Comparison of one-step outflow laboratory method to an in situ method for measuring hydraulic conductivity. - Soil Sci. Soc. Amer. J. 44: 903-907, 1980.

*7212 - JEANRENAUD, E.: Le déficit hydrique sublétal et le degré de résistance à la sécheresse de quelques plantes de divers groupes écologo-physiologique.- An. ştiinţ. Univ. "Al.I. Cuza" Iasi Sect. IIa, Biol. 21: 20-22, 1975.

*7213 - JEANRENAUD, E.: La corrélation entre la résistance à la secheresse et la te-
neur en acides organiges chez des espèces de divers groupes écophysiologiques.
- An. științ. Univ. "Al.I. Cuza" Iași Sect. IIa, Biol. 25; 41-46, 1979.

*7214 - JEANRENAUD, E.: La dynamique des valuers osmotiques chez les espèces de divers
groupes écophysiologiques du littoral de la Mer Noire et leur corrélation avec
le degré de la résistance a la sécheresse. - In: Culegere de Studii și Arti-
cole de Biologie. Vol. 1. Pp. 259-272. Universitatea "Al.I. Cuza",Iași 1979.

*7215 - JEANRENAUD, E., TRUȘCĂ, M., VIDRAȘCU, P.: L'influence de la sécheresse pédo-
logique sur le comportement hydrique et sur le métabolisme glucidique, pro-
téique et des acides organiques chez deux sortes de blé. - An. științ. Univ.
"Al.I. Cuza"Iași Sect. IIa, Biol. 24: 29-32, 1978.

7216 - JEFFERIES, R.L.: The role of organic solutes in osmoregulation in halophytic
higher plants. - In: RAINS, D.W., VALENTINE, R.C., HOLLAENDER, A. (ed.):
Genetic Engineering of Osmoregulation. Impact on Plant Productivity for Food,
Chemicals, and Energy. Pp. 135-154. Plenum Press, New York - London 1980.

7217 - JENNY, H.: The Soil Resource. Origin and Behavior. (Ecological Studies Vol.
37.)- Springer-Verlag, New York - Heidelberg - Berlin 1980.

7218 - JEWER, P.C., INCOLL, L.D.: Promotion of stomatal opening in the grass *Anthe-
phora pubescens* Nees by a range of natural and synthetic cytokinins. - Planta
150: 218-221, 1980.

7219 - JIKA, N.I., ST-PIERRE, C.A., DENIS, J.C.: L'adaptation de cultivars de sorgho-
-grain à différent régimes hydriques. - Can. J. Plant Sci. 60: 233-239, 1980.

*7220 - JINDAL, P.C., SINGH, J.P., GUPTA, O.P.: Effect of salinity on the mineral nu-
trients in mango seedlings. - Ind. J. agr. Sci. 49: 105-109, 1979.

7221 - JOHNSON, D.A.: Improvement of perennial herbaceous plants for drought-stressed
western rangelands. - In: TURNER, N.C., KRAMER, P.J. (ed.): Adaptation of
Plants to Water and High Temperature Stress. Pp. 419-433. John Wiley & Sons,
New York - Chichester - Brisbane - Toronto 1980.

*7222 - JOHNSON, P.S., NORTON, B.E.: Partitioning the soil profile to measure shad-
scale response to water at different depths. - Soil Sci. 128: 121-125, 1979.

7223 - JOHNSON, P.S., NORTON, B.E.: The effects of subsurface irrigation on current
and subsequent year's growth in shadscale. - J. Range Manage. 33: 331-336,
1980.

*7224 - JOHNSON, T., ODIN, H.: Measurements of evapotranspiration using a dynamic ly-
simeter. - Studia forest. Suecica 146: 1-29, 1978.

*7225 - JOHNSSON, A., SKAAR, H.: Alternating perturbation of the water regulatory
system in *Avena* leaves. - Physiol. Plant. 46: 218-220, 1979.

7226 - JONES, C.A., CARABALY, A.: Estimation of leaf water potential in tropical
grasses with the Campbell-Brewster hydraulic press. - Trop. Agr. 57: 305-307,
1980.

7227 - JONES, C.A., PENA, D., CARABALY, A.: Effects of plant water potential, leaf
diffusive resistance, rooting density and water use on the dry matter production of several tropical grasses during short periods of drought stress. -
Trop. Agr. 57: 211-219, 1980.

7228 - JONES, H.G.: Interaction and integration of adaptive responses to water stress:
The implications of an unpredictable environment. - In: TURNER, N.C., KRAMER,
P.J. (ed.): Adaptation of Plants to Water and High Temperature Stress. Pp.
353-365. John Wiley & Sons, New York - Chichester - Brisbane - Toronto 1980.

7229 - JONES, J.W., HESKETH, J.D.: Predicting leaf expansion. - In: HESKETH, J.D., JONES, J.W. (ed.): Predicting Photosynthesis for Ecosystem Models. Vol. II. Pp. 85-122. CRC Press, Boca Raton 1980.

7230 - JONES, J.W., SMAJSTRLA, A.G.: Application of modeling to irrigation management of soybeans. - In: CORBIN, F.T. (ed.): World Soybean Research Conference II: Proceedings. Pp. 571-599. Westview Press, Boulder 1980.

7231 - JONES, M.B., LEAFE, E.L., STILES, W.: Water stress in field-grown perennial ryegrass. I. Its effect on growth, canopy photosynthesis and transpiration. - Ann. appl. Biol. 96: 87-101, 1980.

7232 - JONES, M.B., LEAFE, E.L., STILES, W.: Water stress in field-grown perennial ryegrass. II. Its effect on leaf water status, stomatal resistance and leaf morphology. - Ann. appl. Biol. 96: 103-110, 1980.

7233 - JONES, M.M., OSMOND, C.B., TURNER, N.C.: Accumulation of solutes in leaves of sorghum and sunflower in response to water deficits. - Aust. J. Plant Physiol. 7: 193-205, 1980.

7234 - JONES, M.M., TURNER, N.C.: Osmotic adjustment in expanding and fully expanded leaves of sunflower in response to water deficits. - Aust. J. Plant Physiol. 7: 181-192, 1980.

*7235 - JORDAN, P.W., NOBEL, P.S.: Infrequent establishment of seedings of *Agave deserti* (*Agavaceae*) in the northwestern Sonoran desert. - Amer. J. Bot. 66: 1079-1984, 1979.

7236 - JORDAN, W.R., MILLER, F.R.: Genetic variability in sorghum root systems: Implications for drought tolerance. - In: TURNER, N.C., KRAMER, P.J. (ed.): Adaptation of Plants to Water and High Temperature Stress. Pp. 383-399. John Wiley & Sons, New York - Chichester - Brisbane - Toronto 1980.

*7237 - JOSHI, G.V., KARADGE, B.A.: Effect of sodium chloride on photosynthetic $^{14}CO_2$ assimilation in *Portulaca oleracea* Linn. - Ind. J. exp. Biol. 17: 167-170, 1979.

*7238 - JOVANOVIĆ, R.: Uticaj deficita vlažnosti u zemljištu na morfološka i biološka svojstva hibrida kukuruza različite grupe zrenja. [Effect of water deficit in the soil on the morphological and biological characters of maize hybrids in different maturation groups.] - Savrem. Poljoprivreda 27: 411-417, 1979. [In Croat, ab: E.]

*7239 - JOVANOVIĆ, R., VASIĆ, G.: Zavisnost prinosa kukuruza od nekih agrotehničkih i površine lista u uslovima navodnjavanja. [Effect of some cultural practices and leaf assimilation area on maize yields under irrigated farming conditions.] - Savrem. Poljoprivreda 24: 21-30, 1976. [In Croat, ab: E.]

*7240 - JUNG, J.: Possibilities for optimalization of plant nutrition by new agrochemical substances -- especially in cereals. - In: SCOTT, T.K. (ed.): Plant Regulation and World Agriculture. Pp. 279-307. Plenum Press, New York - London 1979.

7241 - JUNG, P.K., SCOTT, H.D.: Leaf water potential, stomatal resistance, and temperature relations in field-grown soybeans. - Agron. J. 72: 986-990, 1980.

*7242 - JURY, W.A., FRENKEL, H., DEVITT, D., STOLZY, L.H.: Transient changes in the soil-water system from irrigation with saline water: II. Analysis of experimental data. - Soil Sci. Soc. Amer. J. 42: 585-590, 1978.

*7243 - JURY, W.A., FRENKEL, H., STOLZY, L.H.: Transient changes in the soil-water system from irrigation with saline water: I. Theory. - Soil Sci. Soc. Amer. J. 42: 579-585, 1978.

7244 - KAISER, W.J., HORNER, G.M.: Root rot of irrigated lentils in Iran. - Can. J. Bot. 58: 2549-2556, 1980.

7245 - KANDIKO, R.A., TIMMIS, R., WORRALL, J.: Pressure-volume curves of shoots and roots of normal and drought conditioned western hemlock seedlings. - Can. J. Forest Res. 10: 10-16, 1980.

7246 - KAPPEN, L., LANGE, O.L., SCHULZE, E.-D., BUSCHBOM, U., EVENARI, M.: Ecophysiological investigations on lichens of the Negev desert. VII. The influence of the habitat exposure on dew imbibition and photosynthetic productivity. - Flora 169: 216-229, 1980.

7247 - KARAMANOS, A.J.: Response in plant water status to integrated values of soil matric potential calculated from soil water depletion by a field bean crop. - Aust. J. Plant Physiol. 7: 51-66, 1980.

7248 - KARAMI, E., KRIEG, D.R., QUISENBERRY, J.E.: Water relations and carbon-14 assimilation of cotton with different leaf morphology. - Crop Sci. 20: 421-426, 1980.

7249 - KARIMOVA, F.G., GUSEV, N.A.: Vliyanie tsiklicheskogo adenosin-3',5'-monofosfata (tsAMF) na vodoobmen rastitel'nykh kletok. [Effect of cyclic adenosine-3',5'-monophosphate on water exchange in plant cells.] - Fiziol. Rast. 27: 766-772, 1980. [In R, ab: E.]

7250 - KARIMOVA, F.G., GUSEV, N.A., KUZ'MICHEVA, N.V.: Rol' kletochnoĭ obolochki v vodoobmene rastitel'nykh kletok. [Role of cell membrane in water exchange of plant cells.] - Fiziol. Biokhim. kul't. Rast. 12: 285-290, 1980. [In R, ab: E.]

7251 - KARLEN, D.L., ELLIS, R.,Jr., WHITNEY, D.A., GRUNES, D.L.: Influence of soil moisture on soil cation concentration and the tetany potential of wheat forage. - Agron. J. 72: 73-78, 1980.

*7252 - KARMOKER, J.L., VAN STEVENINCK, R.F.M.: The effect of abscisic acid on the uptake and distribution of ions in intact seedlings of Phaseolus vulgaris cv. Redland Pioneer. - Physiol. Plant. 45: 453-459, 1979.

7253 - KARPUSHKIN, L.T., POLYAKOV, M.A., MAKAROV, P.R.: Vlazhnost' vozdukha nad isparyayushcheĭ poverkhnost'yu mezofil'nykh kletok lista. Sposob i rezul'taty opredeleniya. [Air humidity above the evaporating surface of leaf mesophyll cells. A method and results of estimating the humidity.] - Fiziol. Rast. 27: 889-896, 1980. [In R, ab: E.]

7254 - KARTUSCH, B.: Die photosynthetische Leistung zweier Laubbaumarten (Acer platanoides L. und Platanus acerifolia Willd.) im grossstädtischen Biotop. - Flora 170: 114-126, 1980.

*7255 - KATAOKA, H., NAKAGAWA, S., HAYAMA, T., TAZAWA, M.: Ion movements induced by transcellular osmosis in Nitella flexilis. - Protoplasma 99: 179-187, 1979.

*7256 - KATAOKA, K., KANEKO, M.:[Micro-surface structure of leaf blades of cereal crops observed by scanning electron microscope.] - Bull. Fac. Agr. Tamagawa Univ. 17: 26-37, 1977.[In Jap, ab: E.]

7257 - KATSUMI, M., KAZAMA, H., KAWAMURA, N.: Osmotic potential of the epidermal cells of cucumber hypocotyls as affected by gibberellin and cotyledons. - Plant Cell Physiol. 21: 933-937, 1980.

*7258 - KATZ, A., DEHAN, K., ITAI, C.: Kinetic reversal of NaCl effects. - Plant Physiol. 62: 836-837, 1978.

*7259 - KAUFMANN, M.R.: Stomatal control and the development of water deficit in
         Engelmann spruce seedlings during drought. - Can. J. Forest Res. 9: 297-304,
         1979.

*7260 - KAUSS, H., QUADER, H.: *In vitro* activation of a galactosyl transferase involved
         in the osmotic regulation of *Ochromonas*. - Plant Physiol. 58: 295-298, 1976.

*7261 - KAWAMATA, S.: [Relationship between respiration rate of the disordered fruit
         and water balance in leaves on Nijisseiki pear (*Pyrus serotina* var. culta   -
         Rehder).] - Bull. Tokyo agr. exp. Sta. 12: 3-9, 1979. [In Jap, ab: E.]

 7262 - KAWASE, M., WHITMOYER, R.E.: Aerenchyma development in waterlogged plants. -
         Amer. J. Bot. 67: 18-22, 1980.

*7263 - KAZEMI, H., CHAPMAN, S.R., McNEAL, F.H.: Variation in stomatal number in
         spring wheat cultivars. - Cereal Res. Commun. 6: 359-365, 1978.

 7264 - KⁿDREV, T.G., STANKOVA, P.G.: Vliyanie na monometiloviya ester na itakonovata
         kiselina v"rkhu formiraneto na asimilirashchata listna ploshch i chistata
         produktivnost na fotosintezata pri pshenitsata v usloviya na pochveno zasu-
         shavane. [Effect of the itaconic acid monomethyl ester on the formation of
         assimilating leaf surface and net photosynthetic production in wheat under
         conditions of soil water stress.] - Fiziol. Rast. (Sofia) 6(1): 76-82, 1980.
         [In Bulg, ab: E,R.]

 7265 - KEATINGE, J.D.H., GARRETT, M.K., STEWART, R.H.: Response of perennial and
         Italian ryegrass cultivars to temperature and soil water potential. - J. agr.
         Sci. 94: 171-176, 1980.

*7266 - KECK, R.W.: Carbon dioxide exchange rate inhibition in salt stressed soybean.
         - Plant Physiol. 63 (Suppl.) 149, 1979.

 7267 - KEDROWSKI, R.A.: Changes in cold hardiness of introduced and native interior
         Alaskan evergreens in relation to water and lipid content during spring de-
         hardening. - Physiol. Plant. 48: 438-442, 1980.

 7268 - KELLIHER, F.M., KIRKHAM, M.B., TAUER, C.G.: Stomatal resistance, transpira-
         tion, and growth of drought-stressed eastern cottonwood. - Can. J. Forest
         Res. 10: 447-451, 1980.

 7269 - KEMP, P.R., WILLIAMS, G.J.,III: A physiological basis for niche separation
         between *Agropyron smithii* (C_3) and *Bouteloua gracilis* (C_4). - Ecology 61:
         846-858, 1980.

 7270 - KENNEDY, R.A., EASTBURN, J.L., JENSEN, K.G.: C_3 - C_4 photosynthesis in the
         genus *Mollugo*: Structure, physiology and evolution of intermediate characte-
         ristics. - Amer. J. Bot. 67: 1207-1217, 1980.

*7271 - KEUL, M., ANDREI, R., LAZĂR-KEUL, G., VINTILĂ, R.: Acumularea şi efectul
         plumbului şi cadmiului la grîu (*Triticum vulgare*) şi la porumb (*Zea mays*).
         [Accumulation and effect of lead and cadmium in wheat and maize plants.] -
         Stud. Cercetări Biol. 31(1): 49-54, 1979. [In Roum, ab: E.]

 7272 - KHAN, A.A., KARSSEN, C.M.: Induction of secondary dormancy in *Chenopodium
         bonus-henricus* L. seeds by osmotic and high temperature treatments and its
         prevention by light and growth regulators. - Plant Physiol. 66: 175-181,
         1980.

 7273 - KHAN, A.A., PECK, N.H., SAMIMY, C.: Seed osmoconditioning: physiological and
         biochemical changes. - Isr. J. Bot. 29: 133-144, 1980.

*7274 - KHARE, P.K.: Epidermal structure and ontogeny of stomata in *Dipteris wallichii*. - Phytomorphology 28: 400-405, 1978.

7275 - KHAVARI-NEJAD, R.A.: Growth of tomato plants in different oxygen concentrations. - Photosynthetica 14: 326-336, 1980.

*7276 - KING, M.G., RADOSEVICH, S.R.: Tanoak (*Lithocarpus densiflorus*) leaf surface characteristics and absorption of triclopyr. - Weed Sci. 27: 599-604, 1979.

*7277 - KINNANEN, H., SÄKÖ, J.: Irrigation requirements of the strawberry. - Ann. Agr. Fenn. 18: 160-167, 1979.

*7278 - KIRKHAM, M.B.: Plant-water relations and yield of wheat on ridges tilled in the east-west direction. - In: Proceedings of the International Soil Tillage Research Organization ISTRO Vol. 2. Pp. 271-276. University of Hohenheim, Stuttgart 1979.

7279 - KIRKHAM, M.B.: Efficiency of drought-resistant and drought-sensitive wheat cultivars in the use of elements in soil. - Fyton 38: 137-146, 1980.

7280 - KIRKHAM, M.B.: Characteristics of wheat grown with sewage sludge placed at different soil depths. - J. environ. Qual. 9: 13-18, 1980.

7281 - KIRKHAM, M.B.: Movement of cadmium and water in split-root wheat plants. - Soil Sci. 129: 339-344, 1980.

7282 - KIRKHAM, M.B., SMITH, E.L., DHANASOBHON, C., DRAKE, T.I.: Resistance to water loss of winter wheat flag leaves. - Cereal. Res. Commun. 8: 393-400, 1980.

7283 - KIRKPATRICK, J.B., NUNEZ, M.: Vegetation-radiation relationships in mountainous terrain: eucalypt-dominated vegetation in the Risdon Hills, Tasmania. - J. Biogeogr. 7: 197-208, 1980.

*7284 - KIRST, G.O.: Osmotische Adaptation der marinen Planktonalge *Platymonas subcordiformis* (Hazen). - Ber. Deut. bot. Ges. 92: 31-42, 1979.

7285 - KIRST, G.O.: $^{14}CO_2$-fixation in *Valonia utricularis* subjected to osmotic stress. - Plant Sci. Lett. 18: 155-160, 1980.

7286 - KIRST, G.O.: Mannitol accumulation in *Platymonas subcordiformis* after osmotic stresses and the effect of inhibitors. - Z. Pflanzenphysiol. 98: 35-42, 1980.

7287 - KIRST, G.O.: Phosphate transport in *Platymonas subcordiformis* after osmotic stresses. - Z. Pflanzenphysiol. 97: 289-297, 1980.

7288 - KITTLE, D.R., GRAY, L.E.: Effects of infection by *Phytophthora megasperma* var. *sojae* on the water relations of soybean. - Crop Sci. 20: 504-507, 1980.

*7289 - KLENOVSKÁ, S.: Dependence of some physiological processes in tobacco explants upon the carrier of the experimental material. - Acta Fac. Rerum natur. Univ. Comenianae, Ser. Physiol. Plant. 15: 17-23, 1978.

7290 - KLENOVSKÁ, S.: Water relations and the content of metabolites in the course of morphogenetic processes in *Silybum marianum* Gaertn. tissue cultures. - Acta Fac. Rerum natur. Univ. Comenianae, Ser. Physiol. Plant. 17: 31-38, 1980.

*7291 - KLINGAMAN, G.L., LINK, C.B.: Reduction air pollution injury to foliage of *Chrysanthemum morifolium* Ramat. using tolerant cultivars and chemical protectants. - J. Amer. Soc. hort. Sci. 100: 173-175, 1975.

*7292 - KLUGE, M., LORENZEN, H. (ed.): Biochemische Grundlagen Ökologischer Anpassungen bei Pflanzen. - Gustav Fischer Verlag, Stuttgart - New York 1979.

7293 - KNOF, G.: Ein registrierendes Transpirationsporometer zur Messung des Diffu-
sionswiderstandes an Pflanzen in Feldversuchen. - Arch. Acker- Pflanzenbau
Bodenk. 24: 647-654, 1980.

7294 - KNYPL, J.S., JANAS, K.M., RADZIWONOUSKA-JOZWIAK, A.: Is enhanced vigour in
soybean (*Glycine max*) dependent on activation of protein turnover during
controlled hydration of seeds? - Physiol. vég. 18: 157-161, 1980.

*7295 - KOCH, J., BERGMANN, H.: Einfluss von Phytohormonen auf die Wasserausnutzung
(WUE) von *Hordeum vulgare* (L.) und *Triticum aestivum* (L.). - Biochem. Physiol.
Pflanz. 174: 486-490, 1979.

*7296 - KOCH, K.E., KENNEDY, R.A.: Crassulacean acid metabolism (CAM) in *Portulaca
oleracea* L. under natural environmental conditions. - Plant Physiol. 63
(Suppl.): 37, 1979.

7297 - KOCH, K., KENNEDY, R.A.: Characteristics of Crassulacean acid metabolism in
the succulent C_4 dicot, *Portulaca oleracea* L. - Plant Physiol. 65: 193-197,
1980.

7298 - KOHNO, T., SCHMID, M., ROTH, J.R.: Effect of electrolytes on growth of mutant
bacteria. - In: RAINS, D.W., VALENTINE, R.C., HOLLAENDER, A. (ed.): Genetic
Engineering of Osmoregulation. Impact on Plant Productivity for Food, Chemicals, and Energy. Pp. 53-57. Plenum Press, New York - London 1980.

*7299 - KOŁOTA, E., SCIĄŻKO, D.: Wpływ uwilgotnienia gleby i nawożenia mineralnego na
zawartość wapnia i sodu w warzywach. [Effect of the soil moisture level and
the mineral fertilization on the calcium and sodium content in vegetables.] -
Rocz. Nauk roln. Ser. A 104: 115-129, 1979. [In Pol, ab: E,R.]

7300 - KONDO, N., MARUTA, I., SUGAHARA, K.: Effects of sulfite and pH on abscisic
acid-dependent transpiration and on stomatal opening. - Plant Cell Physiol.
21: 817-828, 1980.

*7301 - KÖNIGSHOFER, H., ALBERT, R., KINZEL, H.: Ein ungewöhnliches Zucker-Spektrum
bei *Dianthus lumnitzeri* (Wiesb.). - Z. Pflanzenphysiol. 92: 449-453, 1979.

*7302 - KONONOV, K.E.: Transpiratsiya osnovnykh dominantov poĭmennykh lugov Sredneĭ
Leny. [Transpiration of dominant species of flooded grass swards in middle
Lena region.] - Ėkologiya 1978(4): 20-25, 1978. [In R.]

*7303 - KONTTURI, M.: The effect of weather on yield and development of spring wheat
in Finland. - Ann. Agr. Fenn. 18: 263-274, 1979.

7304 - KÖRNER, C., De MORAES, J.A.P.V.: Water potential and diffusion resistance in
alpine cushion plants on clear summerdays. - Oecol. Plant. 14: 109-120, 1980.

*7305 - KÖRNER, C., SCHEEL, J.A., BAUER, H.: Maximum leaf diffusive conductance in
vascular plants. - Photosynthetica 13: 45-82, 1979.

*7306 - KOSSUTH, S.V., BIGGS, R.H., MARTIN, F.G.: Effect of physiological age of
fruit, temperature, relative humidity, and formulations on absorption of
^{14}C-release by 'Valencia' oranges. - J. Amer. Soc. hort. Sci. 104: 323-327,
1979.

7307 - KOSTŘICA, P., VICHERKOVÁ, M.: Význam draslíku a organických kyselin pro reakce průduchů izolované epidermis bobu. [The role of potassium and organic
acids in stomatal responses of the isolated epidermis of *Vicia faba* L.] -
In: Dny Rostlinné Fyziologie II. Pp. 236-239. Vysoká Škola Zemědělská, Brno
1980. [In Czech, ab: E,R.]

7308 - KOZINKA, V.: Citlivý registračný prístroj pre meranie príjmu vody rastlinami. [A sensitive apparatus to record water uptake by plants.] - In: Dny Rostlinné Fyziologie II. Pp. 240-245. Vysoká Škola Zemědělská, Brno 1980. [In Slov, ab: E,R.]

7309 - KOZINKA, V., SAFWAT MANDOUR, M.: Significance of the primary seminal root in the longitudinal transport of water in the root system of *Sorghum saccharatum* L. (Moench.). - Biológia (Bratislava) 35: 743-752, 1980.

7310 - KOZINKA, V., SLAVÍK, B.: Příjem, transport a výdej vody. [Water uptake, transfer and output in plants.] - In: Dny Rostlinné Fyziologie II. Pp. 41-53. Vysoká Škola Zemědělská, Brno 1980. [In Czech, ab: E,R.]

*7311 - KRAFTI, G., GORANOV, Kh., BANOV, Ĭ., KLEVTSOV, A.: Transpiratsiyata kato komponent na evapotranspiratsiyata. [Transpiration as a component of evapotranspiration.] - Rasteniev"dni Nauki (Sofia) 16(9-10): 49-59, 1979. [In Bulg, ab: R,E.]

7312 - KRAMER, P.J.: Drought, stress, and the origin of adaptations. - In: TURNER, N.C., KRAMER, P.J. (ed.): Adaptation of Plants to Water and High Temperature Stress. Pp. 7-20. John Wiley & Sons, New York - Chichester - Brisbane - Toronto 1980.

7313 - KRATKY, B.A., COX, E.F., McKEE, J.M.T.: Effects of block and soil water content on the establishment of transplanted cauliflower seedlings. - J. hort. Sci. 55: 229-234, 1980.

7314 - KREUZER, H.P., KAUSS, H.: Role of α-galactosidase in osmotic regulation of *Poterioochromonas malhamensis*. - Planta 147: 435-438, 1980.

*7315 - KRISHNAMOORTHY, T.M., SOMAN, S.D.: Incorporation of ^{3}H from HTO exposure in the chemical constituents of algal cells. - Ind. J. exp. Biol. 16: 565-568, 1978.

7316 - KROGMAN, K.K., MACDONALD, M.D., HOBBS, E.H.: Response of silage and grain corn to irrigation and N fertilizer. - Can. J. Plant Sci. 60: 445-451, 1980.

7317 - KROGMAN, K.K., MacKAY, D.C.: Horizon mixing in Solonetzic and associated soils: Effect on drought-stressed barley and wheat. - Can. J. Soil Sci. 60: 721-729, 1980.

7318 - KROGMAN, K.K., McKENZIE, R.C., HOBBS, E.H.: Response of fababean yield, protein production, and water use to irrigation. - Can. J. Plant Sci. 60: 91-96, 1980.

7319 - KU, S.B., EDWARDS, G.E.: Oxygen inhibition of photosynthesis in the C_4 species *Amaranthus graecizans* L. - Planta 147: 277-282, 1980.

*7320 - KUDREV, T., ANDONOVA, P.: Influence of macroelements contents in the nutrient solution of the water regime of maize plants. - In: KUDREV, T., STOYANOV, I., GEORGIEVA, V. (ed.): Mineral Nutrition of Plants. Vol. II. Pp. 245-250. Publishing House, Central Cooperative Union, Sofia 1979.

*7321 - KUENEMAN, E.A., WALLACE, D.H., LUDFORD, P.M.: Photosynthetic measurements of field-grown dry beans and their relation to selection for yield. - J. Amer. Soc. hort. Sci. 104: 480-482, 1979.

7322 - KUHLMAN, E.G.: Influence of moisture on rate of decay of loblolly pine root wood by *Heterobasidion annosum*. - Can. J. Bot. 58: 36-39, 1980.

7323 - KUIPER, P.J.C.: Lipid metabolism of higher plants in saline environments. - Physiol. vég. 18: 83-88, 1980.

7324 - KULICHOVÁ, S.: Naturálna efektívnosť dusíka v závislosti od vodného režimu
       pôdy pri reznačke laločnatej (*Dactylis glomerata* L.). [Natural effectiveness
       of nitrogen in cocksfoot as depending on the soil water regime.] - Rost. Vý-
       roba (Praha) 26: 609-619, 1980. [In Slov, ab: E,R,G.]

*7325 - KUL'TEBAEV, Ė.T.: Vliyanie semennykh podvoev na nakoplenie khlorofillov *a* i *b*
       v list'yakh Aporta Aleksandra. [Effect of stocks on the accumulation of chlo-
       rophylls *a* and *b* in leaves of "Aport Aleksandra".] - Izv. Akad. Nauk Kaz. SSR,
       Ser. biol. 13(4): 21-23, 1975. [In R, ab: Kaz.]

7326 - KUMAKOV, V.A.: Printsipy razrabotki optimal'nykh modeleĭ (ideatipov) sortov
       rasteniĭ. [Principles of optimal variety models (ideotypes) elaboration.] -
       Sel'skokhoz. Biol. 15: 190-197, 1980. [In R, ab: E.]

7327 - KUMMEROW, J.: Adaptation of roots in water-stressed native vegetation. - In:
       TURNER, N.C., KRAMER, P.J. (ed.): Adaptation of Plants to Water and High Tem-
       perature Stress. Pp. 57-73. John Wiley & Sons, New York - Chichester - Bris-
       bane - Toronto 1980.

*7328 - KUNOH, H., KOHNO, M., TASHIRO, S., ISHIZAKI, H.: Studies of the powdery-mildew
       fungus, *Leveillula taurica*, on green pepper. II. Light and electron microsco-
       pic observation of the infection process. - Can. J. Bot. 57: 2501-2508, 1979.

*7329 - KUSHNIRENKO, M.D., KORNESCU, A.S.: Influence of mineral fertilizers and TUR
       preparation on the water regime of apple-trees during irrigation. - In:
       KUDREV, T., STOYANOV, I., GEORGIEVA, V. (ed.): Mineral Nutrition of Plants.
       Vol. I. Pp. 221-226. Publishing House, Central Cooperative Union, Sofia 1979.

*7330 - KUSHNIRENKO, M.D., KRYUKOVA, E.V., PECHERSKAYA, S.N., KANASH, E.V.: Vliyanie
       vodnogo stressa na sostoyanie khloroplastov rasteniĭ razlichnykh ėkologiches-
       kikh grupp. [The effect of water stress on the chloroplasts state in plants
       of different ecological types.] - Izv. Akad. Nauk Mold. SSR, Ser. biol. khim.
       Nauk 1977(3): 17-24, 1977. [In R.]

7331 - KUYAN, V.G.: Vplyv formy krony na fiziologo-biokhimichni protsesy v pagonakh
       i gilkakh yabluni. [Effect of crown shape on physiological and biochemical
       processes in sprouts and branches of apple-tree.] - Ukr. bot. Zh. 37(4):
       49-52, 1980. [In Ukr, ab: E.]

7332 - KUZ'MINA, R.I., KASHKINA, L.V., ABRAMOV, V.L.: O fraktsiyakh vody v zerne kuku-
       ruzy. [Water fractions in maize grains.] - Fiziol. Rast. 27: 266-271, 1980.
       [In R, ab: E.]

*7333 - LACKEY, J.A.: Leaflet anatomy of *Phaseoleae* (*Leguminosae: Papilionoideae*) and
       its relation to taxonomy. - Bot. Gaz. 139: 436-446, 1978.

7334 - LAHIRI, A.N.: Interaction of water stress and mineral nutrition on growth and
       yield. - In: TURNER, N.C., KRAMER, P.J. (ed.): Adaptation of Plants to Water
       and High Temperature Stress. Pp. 341-352. John Wiley & Sons, New York -
       Chichester - Brisbane - Toronto 1980.

7335 - LAISK, A., OJA, V., KULL, K.: Statistical distribution of stomatal apertures
       of *Vicia faba* and *Hordeum vulgare* and the *Spannungsphase* of stomatal opening.
       - J. exp. Bot. 31: 49-58, 1980.

*7336 - LAKHANOV, A.P.: Rol' temperaturnogo faktora v ontogeneze rasteniĭ fasoli.
       [Role of temperature in ontogenesis of *Phaseolus* plants.] - Fiziol. Rast.
       22: 1001-1006, 1975. [In R, ab: E.]

*7337 - LAKSHMINARAYANA, R., PATEL, G.J., JAISANI, B.G.: Note on seed germinability
       in mannitol solution as an index of drought resistance in tobacco. - Ind. J.
       agr. Sci. 49: 818-819, 1979.

☆7338 - LAKSHMINARAYANA, R., PATEL, G.J., JAISANI, B.G.: Gene effects for drought resistance in tobacco. - Ind. J. Genet. Plant Breed. 39: 485-491, 1979.

☆7339 - LAL, B.B., CHAKRAVARTI, B.P.: Factors affecting development of brown spot on maize caused by *Physoderma maydis*. - Iranian J. Plant Pathol. 13(1/2):6-13, 1977.

7340 - LAMOND, M., LEVERT, J.: Influence des enveloppes séminales sur l'imbibition des glands de chêne pédonculé (*Q. robur* L.). - Ann. Sci. forest. 37: 73-83, 1980.

7341 - LANGE, O.L.: Moisture content and CO_2 exchange of lichens I. Influence of temperature on moisture-dependent net photosynthesis and dark respiration in *Ramalina maciformis*. - Oecologia 45: 82-87, 1980.

☆7342 - LAPČEVIĆ, R.: Uticaj broja biljaka i vremena ubiranja na prinos i kvalitet zelene mase nekih hibrida kukuruza. [Effect of plant density and time of harvesting on the yield and quality of the green matter of some maize hybrids.] - Arh. poljopr. Nauke 32: 45-56, 1979. [In Croat, ab: E.]

7343 - LARCHER, W.: Physiological Plant Ecology. 2nd Ed. - Springer-Verlag, Berlin - Heidelberg - New York 1980.

☆7344 - LaROCHE, G.: An experimental study of population differences in leaf morphology of *Aquilegia canadensis* L. (*Ranunculaceae*). - Amer. Midl. Natur. 100: 341-349, 1978.

7345 - LaROCHE, G.: The effects of restricting root growing space, decreasing nutrient supply and increasing water stress on the phenetics of *Aquilegia canadensis* L. (*Ranunculaceae*). - Bull. Torrey bot. Club 107: 220-231, 1980.

☆7346 - LARQUÉ-SAAVEDRA, A., De LEÓN, F.: Stomatal aperture as affected by acetyl salicylic acid (ASA) at different pH's. - Plant Physiol. 63 (Suppl.): 121, 1979.

7347 - LARSON, D.W.: Seasonal change in the pattern of net CO_2 exchange in *Umbilicaria* lichens. - New Phytol. 84: 349-369, 1980.

7348 - LARSON, M.M.: Effects of atmospheric humidity and zonal soil water stress on initial growth of planted northern red oak seedlings. - Can. J. Forest Res. 10: 549-554, 1980.

7349 - LASCÈVE, G., COUCHAT, P.: Le transfert de l'eau dans la plante en régime transitoire. - Ann. agron. 31: 273-283, 1980.

☆ 7350 - LASSOIE, J.P.: Stem dimensional fluctuations in Douglas-fir of different crown classes. - Forest Sci. 25: 132-144, 1979.

☆ 7351 - LASSOIE, J.P., CHAMBERS, J.L.: The effects of an extreme drought on tree water status and net assimilation rates of a transplanted northern red oak under greenhouse conditions. - In: FRALISH, J.S., WEAVER, G.T., SCHLESINGER, R.C. (ed.): Central Harwood Forest Conference. Pp. 269-283. University of Illinois, Carbondale 1976.

☆7352 - LASSOIE, J.P., SCOTT, D.R.M.: Water relations of vine maple in a Douglas-fir stand. - In: Proceedings of the Symposium on Terrestrial and Aquatic Ecological Studies of the Northwest. Pp. 23-37. Eastern Washington State College Press, Cheney 1976.

7353 - LAVENDER, D.P.: Effects of the environment upon the shoot growth of woody plants. - In: LITTLE, C.H.A.: Control of Shoot Growth in Trees. Pp. 76-106. Maritimes Forest Research Centre, Fredericton 1980.

7354 - LAWRENCE, T., KORVEN, H.C., WINKLEMAN, G.E., WARDER, F.G.: The productivity
       and chemical composition of Altai wild ryegrass as influenced by time of
       irrigation and time and rate of N fertilization. - Can. J. Plant Sci. 60:
       1179-1189, 1980.

7355 - LAZAREV, Y.A., TERPUGOV, E.L.: Effect of water on the structure of bacterio-
       rhodopsin and photochemical processes in purple membranes. - Biochim. biophys.
       Acta 590: 324-338, 1980.

*7356 - LEACH, G.J.: The ecology of lucerne pastures. - In: WILSON, J.R. (ed.): Plant
       Relations in Pastures. Pp. 290-308. CSIRO, Melbourne 1978.

7357 - LEACH, J.E.: Photosynthesis and growth of spring barley: some effects of
       drought. - J. agr. Sci. 94: 623-635, 1980.

*7358 - LEE-STADELMANN, O., STADELMANN, E.J.: Drought tolerance and protoplasmic
       qualities in mesophytic higher plants. - In: GOODIN, J.R., NORTHINGTON, D.K.
       (ed.): Arid Land Plant Resources. Pp. 501-528. Texas Technical University,
       Lubbock 1979.

*7359 - LENKA, D., SAHU, S.K.: Effect of water management on quality of crops. - Food
       Farming Agr. 9: 114-116, 1977.

7360 - LERCH, G.: Pflanzenökologie. Teil I: Das Pflanzenleben in Seiner Natürlichen
       Umwelt. Wissenschaftliche Taschenbücher, Reihe Biologie, Bd. 262. - Akademie-
       -Verlag, Berlin 1980.

7361 - LERCH, G.: Pflanzenökologie. Teil II: Zur Ökologie von Stoffproduktion und
       Ertragsbildung. Wissenschaftliche Taschenbücher, Reihe Biologie, Bd. 263. -
       Akademie-Verlag, Berlin 1980.

*7362 - LEVERENZ, J.W., JARVIS, P.G.: Photosynthesis in Sitka spruce. VIII. The effects
       of light flux density and direction on the rate of net photosynthesis and the
       stomatal conductance of needles. - J. appl. Ecol. 16: 919-932, 1979.

*7363 - LEVIN, R.L.: Water permeability of yeast cells at sub-zero temperatures. -
       J. Membrane Biol. 46: 91-112, 115-124, 1979.

7364 - LEVITT, J.: Stress terminology. - In: TURNER, N.C., KRAMER, P.J. (ed.):
       Adaptation of Plants to Water and High Temperature Stress. Pp. 437-439.
       John Wiley & Sons, New York - Chichester - Brisbane - Toronto 1980.

7365 - LEVY, Y.: Effect of evaporative demand on water relations of *Citrus limon*. -
       Ann. Bot. 46: 695-700, 1980.

7366 - LINEBERGER, R.D., STEPONKUS, P.L.: Cryoprotection by glucose, sucrose, and
       raffinose to chloroplast thylakoids. - Plant Physiol. 65: 298-304, 1980.

*7367 - LINTHURST, R.A.: The effect of aeration on the growth of *Spartina alterniflora*
       Loisel. - Amer. J. Bot. 66: 685-691, 1979.

7368 - LIPE, W.N., THOMAS, D.G.: Effect of antitranspirants on yield of Norgold
       Russet potatoes under greenhouse and field conditions. - Amer. Potato J. 57:
       267-273, 1980.

7369 - LIPPS, P.E.: The influence of temperature and water potential on asexual re-
       production by *Pythium* spp. associated with snow rot of wheat. - Phytopatho-
       logy 70: 794-797, 1980.

*7370 - LISTER, R., LEMON, E.: Interactions of atmospheric carbon dioxide, diffuse
       light, plant productivity and climate processes - model predictions. - In:
       ENGELMAN, R.J., SEHMEL, G.A. (ed.): Atmosphere-Surface Exchange of Particulate
       and Gaseous Pollutants. Pp. 112-135. National Technical Information Service,
       Springfield 1976.

*7371 - LLOYD, N.D.H., WOOLHOUSE, H.W.: Comparative aspects of photosynthesis, photo-
respiration and transpiration in four species of the *Cyperaceae* from the re-
lict flora of Teesdale, Northern England. - New Phytol. 83: 1-7, 1979.

7372 - LOIJENS, H.S.: Determination of soil water content from terrestrial gamma
radiation measurements. - Water Resour. Res. 16: 565-573, 1980.

*7373 - LONGDEN, P.C., JOHNSON, M.G., DARBY, R.J., SALTER, P.J.: Establishment and
growth of sugar beet as affected by seed treatment and fluid drilling. - J.
agr. Sci. 93: 541-552, 1979.

*7374 - LONGSTRETH, D.J., NOBEL, P.S.: Nutrient effects on photosynthesis of cotton.
- Plant Physiol. 63(Suppl.): 39, 1979.

7375 - LONGSTRETH, D.J., NOBEL, P.S.: Nutrient influences on leaf photosynthesis.
Effects of nitrogen, phosphorus, and potassium for *Gossypium hirsutum* L. -
Plant Physiol. 65: 541-543, 1980.

*7376 - LONKERD, W.E., RITCHIE, J.T.: Split root observation system for root dynamics
studies. - Agron. J. 71: 519-522, 1979.

*7377 - LOPUSHINSKY, W.: Water relations and photosynthesis in lodgepole pine. - In:
BAUMGARTNER, D.M. (ed.): Management of Lodgepole Pine Ecosystems. Symposium
Proceedings. Pp. 135-153. Washington State University Cooperative Extension
Service, Pullman 1975.

7378 - LOPUSHINSKY, W.: Occurrence of root pressure exudation in pacific northwest
conifer seedlings. - Forest Sci. 26: 275-279, 1980.

7379 - LOPUSHINSKY, W., KLOCK, G.O.: Effect of defoliation on transpiration in grand
fir. - Can. J. Forest Res. 10: 114-116, 1980.

*7380 - LÖSCH, R., BRESSEL, C.: Die Kaliumverteilung in den Schliesszellen leptospo-
rangiater Farne unterschiedlicher Stomatatypen. - Flora 168: 109-120, 1979.

7381 - LOUWERSE, W.: Effects of CO_2 concentration and irradiance on the stomatal behaviour of maize, barley and sunflower plants in the field. - Plant Cell Environ. 3: 391-398, 1980.

*7382 - LOVETT, J.V., SPEAK, M.D.: Studies od *Salvia reflexa* Hornem. II. Examination
of specialized leaf surface structures. - Weed Res. 19: 359-362, 1979.

7383 - LUDLOW, M.M.: Adaptive significance of stomatal responses to water stress. -
In: TURNER, N.C., KRAMER, P.J. (ed.): Adaptation of Plants to Water and High
Temperature Stress. Pp. 123-138. John Wiley & Sons, New York - Chichester -
Brisbane - Toronto 1980.

7384 - LUDLOW, M.M., NG, T.T., FORD, C.W.: Recovery after water stress of leaf gas
exchange in *Panicum maximum* var. *trichoglume*. - Aust. J. Plant Physiol. 7:
299-313, 1980.

7385 - LUGG, D.G., SINCLAIR, T.R.: Seasonal changes in morphology and anatomy of
field-grown soybean leaves. - Crop Sci. 20: 191-196, 1980.

*7386 - LUGO, A.E., GONZALES-LIBOY, J.A., CINTRON, B., DUGGER, K.: Structure, pro-
ductivity, and transpiration of a subtropical dry forest in Puerto Rico. -
Biotropica 10: 278-291, 1978.

7387 - LUKASHEV, E.P., VOZARI, E., KONONENKO, A.A., RUBIN, A.B.: Vliyanie temperatury
i vlazhnosti na élektroindutsirovannyĭ batokhromnyĭ sdvig polosy pogloshche-
niya bakteriorodopsina (Br 570). [Influence of temperature and hydration on
the electric-field-induced bathochromic band shift of bacteriorhodopsin.] -
Biofizika 25: 351-353, 1980. [In R, ab: E.]

*7388 - LURIE, S., HENDRIX, D.L.: Differential ion stimulation of plasmalemma adenosi-
ne triphosphatase from leaf epidermis and mesophyll of *Nicotiana rustica* L. -
Plant Physiol. 63: 936-939, 1979.

7389 - LÜTTGE, U., FISCHER, K.: Light-dependent net CO-evolution by C_3 and C_4 plants.
- Planta 149: 59-63, 1980.

*7390 - LUXOVÁ, M., GAŠPARÍKOVÁ, O., PŠENÁKOVÁ, T., POLERECKÝ, O.: Inhibícia rastu a
akumulácia prolínu v klíčiacich rastlinách imbredných linif kukurice ako reak-
cia na osmotický stres. [Growth inhibition and proline accumulation in germinating plants of inbred maize lines as a response to osmotic stress.] - Rost.
Výroba (Praha) 25: 1215-1224, 1979. [In Slov, ab: E,R,G.]

*7391 - LYNCH, F.J., GEOGHEGAN, M.J.: The role of pigmentation in survival of the leaf
spot fungus *Cercospora beticola*. - Ann. appl. Biol. 91: 313-318, 1979.

7392 - LYNCH, J.J., ELWIN, R.L., MOTTERSHEAD, B.E.: The influence of artificial wind-
breaks on loss of soil water from a continuously grazed pasture during a dry
period. - Aust. J. exp. Agr. anim. Husb. 20: 170-174, 1980.

7393 - LYONS, S.M., GIFFORD, G.F.: Impact of incremental surface soil depths on plant
production, transpiration ratios, and nitrogen mineralization rates. - J.
Range Manage. 33: 189-196, 1980.

*7394 - MAAS, E.V., FINKEL, M.: Origin of proteins released from barley roots by osmo-
tic shock. - Plant Sci. Lett. 17: 7-12, 1979.

*7395 - MAAS, E.V., OGATA, G., FINKEL, M.H.: Salt-induced inhibition of phosphate
transport and release of membrane proteins from barley roots. - Plant Physiol.
64: 139-143, 1979.

*7396 - MacDONALD, J.D., DUNIWAY, J.M.: Use of fluorescent antibodies to study the
survival of *Phytophthora megasperma* and *P. cinnamoni* zoospores in soil. -
Phytopathology 69: 436-441, 1979.

*7397 - MADHUSUDANA RAO, I., SWAMY, P.M., DAS, V.S.R.: The reversal of scotoactive
stomatal behavior in some woody weeds by paraquat and 2,4,5,-T. - Weed Sci.
25: 469-472, 1977,

*7398 - MADHUSUDANA RAO, I., SWAMY, P.M., DAS, V.S.R.: CAM-syndrome in some nonsuccu-
lents and its inhibition by paraquat. - In: COOMBS, J. (ed.): 4th International Congress on Photosynthesis. P. 233. UKISES, London 1977.

7399 - MAGNUSSEN, S.: Wasserhaushaltsuntersuchungen bei unterschiedlich beschatteten
Küstentannen (*Abies grandis* Lindl.). - Allg. Forst- Jagzeit. 151: 227-236,
1980.

7400 - MAHLER, R.L., WOLLUM, A.G.,II: Influence of water potential on the survival
of rhizobia in a Goldsboro loamy sand. - Soil Sci. Soc. Amer. J. 44: 988-992,
1980.

7401 - MAIER, M., KAPPEN, L.: Cellular compartmentalization of salt ions and protec-
tive agents with respect to freezing tolerance of leaves. Investigations with
the halophyte *Halimione portulacoides* (L.) Aellen. - Oecologia 38: 303-316,
1979.

7402 - MAIER-MAERCKER, U.: "Peristomatal transpiration" and stomatal movement: a
controversial view. VI. Lanthanum deposits in the epidermal apoplast. - Z.
Pflanzenphysiol. 100: 121-130, 1980.

7403 - MAIER-MAERCKER, U., JAHNKE, A.: Microautoradiography with ^{43}K: A method for the reliable tracing of ion transport in stomata. - Z. Pflanzenphysiol. 100: 35-42, 1980.

*7404 - MAITI, R.K., JANA, A.K.: Histo-morphological study in X_1 generations of seedlings of mesta and roselle (*Hibiscus cannabinus* L. and *Hibiscus sabdariffa* L. var. Altissima). - Bull. bot. Soc. Bengal. 31: 104-108, 1978.

*7405 - MAJID, M.A., SHAIKH, M.A.Q., BEGUN, S., AHMED, Z.U.: Genotypic variability for frequency, distribution and size of stomata in jute (*Corchorus capsularis* L.). - Beitr. Biol. Pflanz. 54: 399-406, 1978.

7406 - MAJOR, D.J.: Effect of simulated frost injury induced by paraquat on kernel growth and development in corn. - Can. J. Plant Sci. 60: 419-426, 1980.

*7407 - MAKSIMOV, V.M., KOBOZEV, I.V.: Ravnomernost' otrastaniya i produktivnost' bobovogo i bobovo-zlakovogo travostoev pri oroshenii i udobrenii v usloviyakh lesostepi USSR. [Regularities of the growth and productivity of legume and legume-grass stands under irrigation and fertilization in the forest-steppes of the Ukrainian SSR.] - Izv. Timiryazev. sel'skokhoz. Akad. 1979(4): 45-55, 1979. [In R, ab: E.]

7408 - MAKUS, D.J., CHOTENA, M., SIMPSON, W.R., ANDEREGG, J.C.: Water stress and sweet corn seed production. - Idaho agr. exp. Sta. Curr. Inform. Ser. 521: 1-3, 1980.

7409 - MALAKONDAIAH, N., RAJESWARARAO, G.: Effect of foliar application of phosphorus on chlorophyll content, Hill reaction, photophosphorylation and $^{14}CO_2$ fixation under salt stress in peanut plants. - Photosynthetica 14: 17-21, 1980.

7410 - MALIK, D.S., OSWAL, M.C., BAKSHI, R.K.: Interaction of profile soil moisture and row spacing on the performance of chickpea on drylands. - Ind. J. agr. Sci. 50: 698-700, 1980.

*7411 - MALOFEEV, V.M., AFANAS'EV, V.P., SHURAVILIN, A.V.: Sravnitel'no-fiziologicheskie issledovaniya khlopchatnika pri oroshenii vodami razlichnoǐ stepeni solenosti. [Comparative physiological researches of the cotton-plant irrigated with waters containing various amounts of salt.] - Sel'skokhoz. Biol. 6: 767-770, 1979. [In R, ab: E.]

7412 - MANN, J.D., EDGE, E.A., LANCASTER, J.E., BLYTH, K.: Growth and solasodine production by *Solanum aviculare* and *Solanum laciniatum* under moisture stress at different temperatures. - N.Zeal. J. agr. Res. 23: 361-366, 1980.

7413 - MANSFIELD, D.H., ANDERSON, J.E.: Measuring plant water status: A simple method for investigative laboratories. - Amer. Biol. Teacher 42: 541-544, 1980.

7414 - MANUǏL'S'KYǏ, V.D., MARTYN, G.G., SYTNYK, K.M.: Kriorezystentnist' pylku roslyn pry rizniǐ vologosti. [Cryoresistance of plant pollen at different moisture.] - Ukr. bot. Zh. 37(4): 14-15, 1980. [In Ukr, ab: E.]

7415 - MAOTANI, T., MACHIDA, Y.: [Leaf water potential as an indicator of irrigation timing for satsuma mandarin trees in summer.] - J. Jap. Soc. hort. Sci. 49: 41-48, 1980. [In Jap, ab: E.]

*7416 - MARK, A.F., ROWLEY, J.: Water yield of low-alpine snow tussock grassland in Central Otago. - J. Hydrol. (N. Zeal.) 15: 59-79, 1976.

*7417 - MARKHART, A.H.,III, FISCUS, E.L., NAYLOR, A.W., KRAMER, P.J.: Effect of abscisic acid on root hydraulic conductivity. - Plant Physiol. 64: 611-614, 1979.

7418 - MARKHART, A.H., III, NAYLOR, A.W.: Effect of valinomycin and gramicidin D on the reflection coefficient of soybean root systems. - Plant Physiol. 65: 74-77, 1980.

7419 - MARKHART, A.H., III, PEET, M.M., SIONIT, N., KRAMER, P.J.: Low temperature acclimation of root fatty acid composition, leaf water potential, gas exchange and growth of soybean seedlings. - Plant Cell Environ. 3: 435-441, 1980.

7420 - MARSHALL, B., BISCOE, P.V.: A model for C_3 leaves describing the dependence of net photosynthesis on irradiance. II. Application to the analysis of flag leaf photosynthesis. - J. exp. Bot. 31: 41-48, 1980.

7421 - MARSHALL, B., SEDGLEY, R.H., BISCOE, P.V.: Effects of a water stress on the photosynthesis and respiration of wheat ears. - Aust. J. agr. Res. 31: 857-871, 1980.

7422 - MARTENS, J., FRETZ, T.A.: Differentiation of nine crabapples based on bud and leaf surface features. - J. Amer. Soc. hort. Sci. 105: 263-273, 1980.

7423 - MASON, W.K., CONSTABLE, G.A., SMITH, R.C.G.: Irrigation for crops in a sub-humid environment. II. The water requirements of soybeans. - Irrig. Sci. 2: 13-22, 1980.

7424 - MASON, W.K., TAYLOR, H.M., BENNIE, A.T.P., ROWSE, H.R., REICOSKY, D.C., JUNG, Y., RIGHES, A.A., YANG, R.L., KASPAR, T.C., STONE, J.A.: Soybean Row Spacing and Soil Water Supply: Their Effect on Growth, Development, Water Relations, and Mineral Uptake. - U.S. Department of Agriculture, Peoria 1980.

7425 - MASSIMINO, D., ANDRÉ, M., RICHAUD, C., DAGUENET, A., MASSIMINO, J., VIVOLI, J.: Évolution horaire au cours d'une journée normale de la photosynthèse, de la transpiration, de la respiration foliaire et racinaire et de la nutrition N.P.K. chez Zea mays. - Physiol. Plant. 48: 512-518, 1980.

7426 - MASSMAN, W.J.: Water storage on forest foliage: A general model. - Water Resour. Res. 16: 210-216, 1980.

7427 - MASZKIEWICZ, J., BLASZCZAK, W., MILLIKAN, D.F.: Changes in phloridzin content, osmotic values of cellular sap, and cell wall thickness of apple leaf tissue associated with proliferation disease. - Phytopathol. Z. 99: 33-36, 1980.

*7428 - MATHUR, D.D., HENDERSHOTT, C.H., VINES, H.M.: Efficiency of water utilization in Crassulacean acid metabolism plants when IN CAM versus OUT OF CAM. - Ind. J. Plant Physiol. 21: 7-11, 1978.

*7429 - MAWSON, B.T., CUMMINS, W.R., COLMAN, B.: The effect of abscisic acid on photosynthesis of isolated mesophyll cells. - Plant Physiol. 63 (Suppl.): 80, 1979.

*7430 - MAYEUX, H.S., Jr., SCIFRES, C.J.: Vegetative growth and mortality of Drummond's goldenweed (Isocoma drummondii [T. & G.] Greene: Compositae) as influenced by soil moisture and site characteristics. - Southwest. Natur. 24: 97-114, 1979.

7431 - MAYO, J.M., EHRET, D.: The effects of abscisic acid and vapor pressure deficit on leaf resistance of Paphiopedilum leeanum. - Can. J. Bot. 58: 1202-1204, 1980.

*7432 - McANENEY, K.J., TANNER, C.B., GARDNER, W.R.: An in-situ dewpoint hygrometer for soil water potential measurement. - Soil Sci. Soc. Amer. J. 43: 641-645, 1979.

7433 - McGOWAN, M., WILLIAMS, J.B.: The water balance of an agricultural catchment. I. Estimation of evaporation from soil water records. - J. Soil Sci. 31: 217-230, 1980.

7434 - McGOWAN, M., WILLIAMS, J.B.: The water balance of an agricultural catchment
II. Crop evaporation: Seasonal and soil factors. - J. Soil Sci. 31: 231-244,
1980.

7435 - McGOWAN, M., WILLIAMS, J.B., MONTEITH, J.L.: The water balance of an agricultural catchment III. The water balance. - J. Soil Sci. 31: 245-262, 1980.

7436 - McHUGH, J.J.,Jr., NISHIMOTO, R.K.: Effect of overhead sprinkler irrigation on
watercress yield, quality, and leaf temperature. - HortScience 15: 801-802,
1980.

*7437 - McINTYRE, G.I.: Developmental studies on *Euphorbia esula*. Evidence of competition for water as a factor in the mechanism of root bud inhibition. - Can. J.
Bot. 57: 2572-2581, 1979.

7438 - McINTYRE, G.I.: The role of water distribution in plant tropisms. - Aust. J.
Plant Physiol. 7: 401-413, 1980.

7439 - McKERSIE, B.D., STINSON, R.H.: Effect of dehydration on leakage and membrane
structure in *Lotus corniculatus* L. seeds. - Plant Physiol. 66: 316-320, 1980.

7440 - McKERSIE, B.D., TOMES, D.T.: Effects of dehydration treatments on germination,
seedling vigour, and cytoplasmic leakage in wild oats and birdsfoot trefoil. -
Can. J. Bot. 58: 471-476, 1980.

*7441 - MEDYANNIKOV, V.M., KHOLUPENKO, I.P.: Postuplenie ^{14}C-assimilyatov v list'ya
soi v zavisimosti ot udel'noĭ radioaktivnosti ^{14}CO$_2$ i zavyadaniya rasteniĭ.
[Transport of ^{14}C-photosynthates into soybean leaves in dependence on specific
radioactivity of ^{14}CO$_2$ and plant wilting.] - In: Pogloshchenie i Peredvizhenie
Veshchestv u Rasteniĭ. Pp. 35-38. Dal'nevostoch. nauch. Tsentr Akad. Nauk
SSSR, Biol.-Pochven. Inst., Vladivostok 1978. [In R.]

7442 - MEEK, B.D., OWEN-BARTLETT, E.C., STOLZY, L.H., LABANAUSKAS, C.K.: Cotton yield
and nutrient uptake in relation to water table depth. - Soil Sci. Soc. Amer.
J. 44: 301-305, 1980.

*7443 - MEENAKSHI, R.M., GNANARETHINAM, J.L.: Phytochemical aspects of onion (*Allium
cepa*, Linn.) under water stress. - Plant Physiol. 63 (Suppl.): 89, 1979.

7444 - MEETEREN, U., van: Water relations and keeping-quality of cut Gerbera flowers.
VI. Role of pressure potential. - Scientia Hort. 12: 283-292, 1980.

7445 - MEETEREN, U., van, GELDER, H., van: Water relations and keeping-quality of cut
Gerbera flowers. V. Role of endogenous cytokinins. - Scientia Hort. 12: 273-
281, 1980.

7446 - MEIERING, A.G., PAROSCHY, J.H., PETERSON, R.L., HOSTETTER, G., NEFF, A.: Mechanical freezing injury in grapevine trunks. - Amer. J. Enol. Viticult. 31:
81-89, 1980.

*7447 - MELEKHOV, E.I.: Vliyanie 2,4-D na provodimost' korneĭ, pogloshchnie vody i
transpiratsiyu. [Effect of 2,4-D on root conductivity, water uptake and transpiration.] - Fiziol. Rast. 26: 1001-1007, 1979. [In R, ab: E.]

*7448 - MENEZES, E.M., de, STAFFUZA, E.M.: Estudo sobre a folha de *Dorstenia bryonii-
folia* Mart. ex Miq. (*Moraceae, Moroideae*). [Studies on the leaf of *Dorstenia
bryoniifolia* Mart. ex Miq. (*Moraceae, Moroideae*).] - Rev. Brasil. Biol. 39:
291-303, 1979. [In: Portug., ab: E.]

7449 - MENOUX-BOYER, Y.: Effets et post-effets d'une sécheresse édaphique temporaire
modérée sur la croissance du lin: Influence des conditions énergétiques du
milieu. - Acta Oecol. - Oecol. Plant. 1: 55-69, 1980.

7450 - **MENYAÏLO, L.N.**: Vliyanie vlazhnosti pochvy na rost i soderzhanie gormonal'nykh veshchestv v khvoe sosny. [Influence of soil humidity on growth and the content of hormonal substances in pine needles.] - Fiziol. Rast. 27: 639-641, 1980. [In R.]

7451 - **MERIAUX, S.**: Effects of climate on the yield response of two fescue species to water and nitrogen application. - Irrig. Sci. 1: 233-239, 1980.

*7452 - **MÉRIAUX, S., ROLLIN, H., RUTTEN, P.**: Effets de la sécheresse sur la vigne. I. - Etudes sur Cabernet-Sauvignon. - Ann. agron. 30: 553-575, 1979.

*7453 - **MÉRIDA, T., ARIAS, I.**: Estudios fisio-ecologicos en plantas de las zonas aridas y semiaridas de Venezuela. III. Desarrollo de gloquideas por efecto de acido abscisico (ABA) en plantulas de *Cereus griseus* Haw. (*Cactaceae*). [Ecophysiological studies of plant species from the arid and semi-arid regions of Venezuela. III. Effects of abscisic acid (ABA) in spines development in seedlings of *Cereus griseus* Haw. (*Cactaceae*).]- Acta Cient. Venezolana 30: 162-166, 1979. [In Span, ab: E.]

*7454 - **MERMERSKA, E., BOÏKOV, S.**: Napoyavane na zh"ltiya mak (*Glaucium flavum* Crantz var. *lejocarpum* Boiss) v Kraïdunavskata nizina na Severozapadna B"lgariya. I. Poliven rezhim i nachini na napoyavane. [Irrigation of yellow horn poppy (*Glaucium flavum* Crantz var. *lejocarpum* Boiss) grown in the Danube plain of Northwestern Bulgaria. I. Irrigation regime and irrigation types.] - Rasteniev. Nauki 16(8): 70-77, 1979. [In Bulg, ab: R, E.]

*7455 - **MERMERSKA, E., BOÏKOV, S.**: Napoyavane na zh"ltiya mak (*Glaucium flavum.* Crantz var. *lejocarpum* Boiss.) v Kraïdunavskata nizina na Severozapadna B"lgariya. II. Vodorazhod. [Irrigation of yellow horn poppy (*Glaucium flavum.* Crantz var. *lejocarpum* Boiss.) grown in the Danube plain of Northwestern Bulgaria. II. Evapotranspiration.] - Rasteniev. Nauki 16(9-10): 65-71, 1979. [In Bulg, ab: R, E.]

7456 - **METOCHIS, C.**: Irrigation of lucerne under semi-arid conditions in Cyprus. - Irrig. Sci. 1: 247-252, 1980.

7457 - **MEYER, W.S., GREEN, G.C.**: Water use by wheat and plant indicators of available soil water. - Agron. J. 72: 253-257, 1980.

7458 - **MEYER, W.S., RITCHIE, J.T.**: Resistance to water flow in the sorghum plant. - Physiol. Plant. 65: 33-39, 1980.

7459 - **MEYER, W.S., RITCHIE, J.T.**: Water status of cotton as related to taproot length. - Agron. J. 72: 577-580, 1980.

*7460 - **MICHAEL, D., DICKMANN, D., NELSON, N.**: Photosynthesis, CO_2 compensation and stomatal conductance of young poplar plants grown under intensive culture. - Plant Physiol. 63 (Suppl.): 121, 1979.

7461 - **MIDMORE, D.J., McDAVID, C.R.**: Factors affecting diurnal trends of ^{14}C fixation in sugar-cane. - Trop. Agr. 57: 203-209, 1980.

7462 - **MIEDEMA, P., SINNAEVE, J.**: Photosynthesis and respiration of maize seedlings at suboptimal temperatures. - J. exp. Bot. 31: 813-819, 1980.

*7463 - **MIGNUCCI, J.S., BOYER, J.S.**: Inhibition of photosynthesis and transpiration in soybean infected by *Microsphaera diffusa*. - Phytopathology 69: 227-230, 1979.

7464 - **MIGUS, W.N., HUNT, L.A.**: Gas exchange rates and nitrogen concentrations in two winter wheat cultivars during the grain-filling period. - Can. J. Bot. 58: 2110 - 2116, 1980.

7465 - **MIKHALEVSKAYA, O.B.**: Izmenenie vodouderzhivayushcheĭ sposobnosti v ontogeneze
       list'ev. [Changes in water-retaining capacity during leaf ontogenesis.] - Fi-
       ziol. Rast. 27: 880-883, 1980. [In R.]

7466 - **MILBORROW, B.V.**: A distinction between the fast and slow responses to abscisic
       acid. - Aust. J. Plant Physiol. 7: 749-754, 1980.

7467 - **MILBORROW, B.V.**: Regulation of abscisic acid metabolism. - In: SKOOG, F.
       (ed.): Plant Growth Substances 1979. Pp. 262-273. Springer-Verlag, Berlin -
       Heidelberg - New York 1980.

*7468 - **MILBURN, J.A.**: An ideal viscous flow porometer. - J. exp. Bot. 30: 1021-1034,
       1979.

7469 - **MILBURN, J.A.**: The measurement of turgor pressure in sieve tubes. - Ber. Deut.
       bot. Ges. 93: 153-166, 1980.

7470 - **MILIČ, M., SPASOJEVIČ, M.**: Zavisnost izmedu evaporacije i evapotranspiracije
       kukuruza na lizimetarskoj stanici u Prizrenu. [Dependence between evaporation
       and evapotranspiration of maize at the lysimeter research station in Prizren.]
       - Arhiv poljopr. Nauke 41: 577-590, 1980. [In Croat, ab: E.]

*7471 - **MILICĂ, C.I.**: Fiziologia porumbului in condiții dirijate de nutriție si umidi-
       tate a solului. [Corn physiology within controlled conditions of nutrition and
       soil moisture.] - Lucrări știint. Ser. Agron. 1978: 45-46, 1978. [In Roum,
       ab: E.]

*7472 - **MILICĂ, C.I., IACOB, T.**: Procesele fiziologice și biochimice la lucernă în
       regim dirijat de irigare și fertilizare. [Physiological and biochemical pro-
       cesses within controlled irrigation and fertilization regime in alfalfa.] -
       Lucrări știint. Ser. Agron. 1978: 91-93, 1978. [In Roum, ab: E.]

7473 - **MILLER, D.E., HANG, A.N.**: Deficit, high-frequency irrigation of sugarbeets
       with the line source technique. - Soil Sci. Soc. Amer. J. 44: 1295-1298, 1980.

7474 - **MILLER, D.R.**: The two-dimensional energy budget of a forest edge with field
       measurements at a forest-parking lot interface. - Agr. Meteorol. 22: 53-78,
       1980.

7475 - **MILLER, D.R., VAVRINA, C.A., CHRISTENSEN, T.W.**: Measurement of sap flow and
       transpiration in ring-porous oaks using a heat pulse velocity technique. -
       Forest Sci. 26: 485-494, 1980.

7476 - **MILLER, N.A.**: The effect of hexadecanol-octadecanol on the leaf water balance
       of *Zea mays* L. - Bot. Gaz. 141: 192-194, 1980

7477 - **MIRHADI, M.J., KOBAYASHI, Y.**: Studies on the productivity of grain sorghum
       III. Comparative investigation of the effect of wilting treatments and foliar
       spray applications of NAA, IAA and tryptophan on grain and forage yields of
       grain sorghum. - Jap. J. Crop Sci. 49: 445-455, 1980.

*7478 - **MISHNEVA, G.F.**: Vodnyĭ rezhim mezhvidovykh privivok sosny v svyazi s ikh ros-
       tom i razvitiem. [The water regime of interspecific inoculations of pine trees
       in relation with their growth and development.] - In: Voprosy Gornogo Lesove-
       deniya v Gruzii. Vol. 25. Pp. 175-180. Sabchota Adzhara, Matumi 1976. [In R,
       ab: E.]

7479 - **MISRA, R.K., NAGARAJARAO, Y.**: Soil water depletion and leaf water status of
       different varieties of wheat under rainfed conditions. - Ind. J. Agron. 25:
       540-542, 1980.

*7480 - **MITCHELL, K.J., ROBOTHAM, R., WARRINGTON, I.**: Physics of controlled environ-
       ment and plant growth. - In: Proceedings of the Symposium on Climate and Rice.
       Pp. 141-155, The International Rice Research Institute, Los Banos 1976.

7481 - MOBAYEN, R.G.: Germination of trifoliate orange seed in relation to fruit de-
       velopment, storage and drying. - J. hort. Sci. 55: 285-289, 1980.

7482 - MOBAYEN, R.G., MILTHORPE, F.L.: Response of seedlings of three citrus-root-
       stock cultivars to salinity. - Aust. J. agr. Res. 31: 117-124, 1980.

7483 - MOGENSEN, V.O.: Drought sensitivity at various growth stages of barley in re-
       lation to relative evapotranspiration and water stress. - Agron. J. 72: 1033-
       1038, 1980.

7484 - MOLDAU, Kh.A., SYBER, Ya.Kh., RAKHI, M.O.: Komponenty temnovogo dykhaniya fa-
       soli pri defitsite vody. [Components of dark respiration of bean plants under
       water stress.] - Fiziol. Rast. 27: 5-10, 1980. [In R, ab: E.]

7485 - MONSELISE, S.P., LENZ, F.: Effects of fruit load on stomatal resistance, spe-
       cific leaf weight, and water content of apple leaves. - Gartenbauwissenschaft
       45: 188-191, 1980.

7486 - MOONEY, H.A.: Seasonality and gradients in the study of stress adaptation. -
       In: TURNER, N.C., KRAMER, P.J. (ed.): Adaptation of Plants to Water and High
       Temperature Stress. Pp. 279-294. John Wiley & Sons, New York - Chichester -
       Brisbane - Toronto 1980.

7487 - MOONEY, H.A., GULMON, S.L., EHLERINGER, J., RUNDEL, P.W.: Atmospheric water
       uptake by an Atacama Desert shrub. - Science 209: 693-694, 1980.

7488 - MOONEY, H.A., GULMON, S.L., RUNDEL, P.W., EHLERINGER, J.: Further observations
       on the water relations of Prosopis tamarugo of the northern Atacama desert. -
       Oecologia 44: 177-180, 1980.

*7489 - MOORE, H.E.,Jr.: The genus Hyophorbe (Palmae). - Gentes Herbarum 11: 212-
       245, 1978.

7490 - MORE HERRERO, A., BETHENOD, O., MOROT-GAUDRY, J.-F.: Influence de la tempéra-
       ture sur les paramètres caractéristiques de la photosynthèse chez de jeunes
       plants de Maïs (var. W64A). - Physiol. vég. 18: 301-312, 1980.

*7491 - MORESHET, S., GREEN, G.C.: Stomatal development and gas exchange in citrus
       fruit in relation to soil moisture. - Plant Physiol. 63 (Suppl.): 122, 1979.

7492 - MORESHET, S., GREEN, G.C.: Photosynthesis and diffusion conductance of the
       Valencia orange fruit under field conditions. - J. exp. Bot. 31: 15-27, 1980.

7493 - MOREY, R.V., GILLEY, J.R., BERGSRUD, F.G., DIRKZWAGER, L.R.: Yield response of
       corn related to soil moisture. - Trans. ASAE 23: 1165-1170, 1980.

7494 - MORGAN, J.M.: Osmotic adjustment in the spikelets and leaves of wheat. - J.
       exp. Bot. 31: 655-665, 1980.

7495 - MORGAN, J.M.: Differences in adaptation to water stress within crop species.
       - In: TURNER, N.C., KRAMER, P.J. (ed.): Adaptation of Plants to Water and
       High Temperature Stress. Pp. 369-382. John Wiley & Sons, New York - Chichester
       - Brisbane - Toronto 1980.

7496 - MORGAN, T.H., BIERE, A.W., KANEMASU, E.T.: A dynamic model of corn yield re-
       sponse to water. - Water Resour. Res. 16: 59-64, 1980.

*7497 - MOTHA, R.P., VERMA, S.B., ROSENBERG, N.J.: Exchange coefficients under sensi-
       ble heat advection determined by eddy correlation. - Agr. Meteorol. 20: 273-
       280, 1979.

*7498 - MOZHEĬKO, G.A.: O meliorativnom vozdeistvii zashchitnykh lesnykh nasazhdeniĭ
       v sukhoĭ stepi USSR. [On the land-improvement action of protective forest
       plantations in the dry steppe of Ukraine.] - Lesovedenie 1978 (1): 21-26,
       1978. [In R, ab: E.]

*7499 - MUCHOW, R.C.: Effects of plant population and season on kenaf (*Hibiscus canna-binus* L.) grown under irrigation in tropical Australia. I. Influence on the components of yield. - Field Crops Res. 2: 55-66, 1979.

*7500 - MUCHOW, R.C.: Effects of plant population and season on kenaf (*Hibiscus canna-binus* L.) grown under irrigation in tropical Australia. II. Influence on growth parameters and yield prediction. - Field Crops Res. 2: 67-76, 1979.

7501 - MUCHOW, R.C., FISHER, M.J., LUDLOW, M.M., MYERS, R.J.K.: Stomatal behaviour of kenaf and sorghum in a semiarid tropical environment. II. During the day. - Aust. J. Plant Physiol. 7: 621-628, 1980.

7502 - MUCHOW, R.C., LUDLOW, M.M., FISHER, M.J., MYERS, R.J.K.: Stomatal behaviour of kenaf and sorghum in a semiarid tropical environment. I. During the night. - Aust. J. Plant Physiol. 7: 609-619, 1980.

7503 - MUCHOW, R.C., WOOD, I.M.: Yield and growth responses of kenaf (*Hibiscus canna-binus* L.) in a semi-arid tropical environment to irrigation regimes based on leaf water potential. - Irrig. Sci. 1: 209-222, 1980.

*7504 - MULLER, R.N., MILLER, J.E., SPRUGEL, D.G.: Photosynthetic response of field--grown soybeans to fumigations with sulphur dioxide. - J. appl. Ecol. 16: 567-576, 1979.

*7505 - MUMFORD, P.M., GROUT, W.W.: Desiccation and low temperature (-196°C) tolerance of *Citrus limon* seed. - Seed Sci. Technol. 7: 407-410, 1979.

7506 - MURASE, H., MERVA, G.E., SEGERLIND, L.J.: Variation of Young's modulus of potato as a function of water potential. - Trans ASAE 23: 794-796, 800, 1980.

7507 - MURATA, M., ROOS, E.E., TSUCHIYA, T.: Mitotic delay in root tips of peas induced by artificial seed aging. - Bot. Gaz. 141: 19-23, 1980.

*7508 - MURRAY, R.B., MAYLAND, H.F., VAN SOEST, P.J.: Seasonal changes in nutritional quality of *Agropyron desertorum* compared with six other semi-arid grasses. - In: GOODIN, J.R., NORTHINGTON, D.K. (ed.): Arid Land Plant Resources. Pp. 538-549. Texas Technical University, Lubbock 1979.

7509 - MUSICK, J.T., DUSEK, D.A.: Planting date and water deficit effects on development and yield of irrigated winter wheat. - Agron. J. 72: 45-52, 1980.

7510 - MYRONYUK, V.I., MASYUK, N.P., AKOPYANTS, N.S.: Vplyv osmotychno diyuchykh spoluk na aktyvnist' katalazy oligo- ta gipergalobnykh vodorosleĩ. [Effect of osmotically effective compounds on the catalase activity of oligo- and hyper-halotic algae.] - Ukr. bot. Zh. 37 (4): 38-42, 1980. [In Ukr, ab: E.]

7511 - NAGARAJAN, S., PANDA, B.C.: A simple laboratory technique for measurement of bound water in wheat leaves. - Ind. J. exp. Biol. 18: 188-191, 1980.

7512 - NAGARAJRAO, Y., MALLICK, S., SINGH, G.: Moisture depletion and root growth of different varieties of chickpea under rainfed conditions. - Ind. J. Agron. 25: 289-293, 1980.

7513 - NAGENDRAN, C.R., AREKAL, G.D., SWAMY, B.G.L.: Facultative stomata in *Griffithe-la (Podostemaceae)*. - Curr. Sci. 49 (14): 561, 1980.

7514 - NAKAMURA, S., ENOHARA, N.: [Germination improvement of vegetable seeds using polyethylene glycol. I. Eggplant, *Cryptotaenia japonica* and carrot.] - J. Jap. Soc. hort. Sci. 48: 443-452, 1980. [In Jap. ab: E.]

*7515 - NAKANO, Y., KAMIGATAGUCHI, Y.: [A simulation model to estimate evapotranspiration from a soybean canopy.] - Sci. Bull. Fac. Agr. Kyushu Univ. 33: 197-207, 1979. [In Jap, ab: E.]

7516 - **NAKAYAMA, A.**: [Differentiating process of stomata in tea leaves.]- Jap. J. Crop Sci. 49: 281-285, 1980. [In Jap, ab: E.]

*7517 - **NAKOS, G.**: Lime-induced chlorosis in *Pinus radiata*. - Plant Soil 52: 527-536, 1979.

*7518 - **NAMBIAR, E.K.S., BOWEN, G.D., SANDS, R.**: Root regeneration and plant water status of *Pinus radiata* D. Don seedlings transplanted to different soil temperatures. - J. exp. Bot. 30: 1119-1131, 1979.

7519 - **NAPP-ZINN, K., EBLE, M.**: Beiträge zur systematischen Anatomie der *Asteraceae-Anthemideae*: Die Trichome. - Plant Syst. Evol. 136: 169-207, 1980.

7520 - **NAPP-ZINN, K., FRANZ, A.**: Vergleichend-anatomische Untersuchungen an petaloiden Hochblättern von *Heliconia revoluta* (Griggs') Standley (Vorläufige Mitteilung). - Flora 170: 565-569, 1980.

*7521 - **NAPP-ZINN, K., HEINS, W.**: Vergleichend-Anatomische Untersuchungen an Petaloiden Hochblättern. II. Acanthaceen. (Tropische und Subtropische Pflanzenwelt 24.) - Akademie der Wiessenschaften und der Literatur, Mainz 1979.

*7522 - **NAPP-ZINN, K., SCHMIDT, R., GENSCHER, H.**: Vergleichend-Anatomische Untersuchungen an Petaloiden Hochblättern. I. Bromeliaceen (Zugleich Abhandlung III der Reihe "Bromelienstudien"). (Tropische und Subtropische Pflanzenwelt 24.) - Akademie der Wiessenschaften und der Literatur, Mainz 1978.

*7523 - **NASSERY, H.**: Salt-induced loss of potassium from plant roots. - New Phytol. 83: 23-27, 1979.

7524 - **NAUTIYAL, S.: PUROHIT, A.N.**: High altitude acclimatization in plants: Stomatal frequency and anatomical changes in leaves of *Artemisia* species. - Biol. Plant. 22: 282-286, 1980.

7525 - **NAVARA, J.**: Die Veränderungen des Wassergehaltes und der Transpirationsintensität bei der Blättern von *Prunus armeniaca* L. unter den Immisionsbedigungen. - Biológia (Bratislava) 35: 17-26, 1980.

7526 - **NEALES, T.F., SALE, P.J.M., MEYER, C.P.**: Carbon dioxide assimilation by pineapple plants, *Ananas comosus* (L.) Merr. II Effects of variation of the day/night temperature regime. - Aust. J. Plant Physiol. 7: 375-385, 1980.

7527 - **NEAT SIMONELLI, M.L., SPOMER, L.A.**: Preparation of customized pressure chamber seals for irregularly-shaped, succulent organs. - Agron. J. 72: 699-700, 1980.

7528 - **NECHIPORENKO, G.A., RYBALOVA, B.A.**: Primenimost' razlichnykh osmoticheski deĭstvuyushchikh agentov dlya issledovaniya vodnogo obmena rastitel'nykh tkaneĭ. [Applicability of various osmotic agents for studying water relations in plant tissues.] - Fiziol. Rast. 27: 203-208, 1980. [In R, ab: E.]

7529 - **NEDBAL, A.F.**: Rezhim orosheniya pertsa sladkogo v Krymu. [Irrigation regime of pepper in Krym.] - Nauch.-tekh. Byull. (Khar'kov) 1980 (12): 13-17, 1980. [In R.]

*7530 - **NEGISI, K.**: Seasonal changes in rate of photosynthesis and growth of *Pinus densiflora, Cryptomeria japonica* and *Chamaecyparis obtusa* seedlings in their second vegetation season. - In: Bicentenary Celebration of C.P. Thunberg's Visit to Japan. Pp. 77-89, Royal Swedish Embassy, Botanical Society of Japan, Tokyo, 1977.

*7531 - **NEIDHART, H.V.**: Ultrastructural changes of the plasmalemma surface during desiccation of *Funaria hydrometrica* spores. - Flora 167: 445-450, 1978.

*7532 - **NELSON, D.C., STEGMAN, E.C.**: Performance of potatoes under irrigation at Oakes, North Dakota. - Farm Res. 32 (5): 21-27, 1975.

7533 - NEWBAUER, J.J., III, WHITE, L.M., MOY, R.M., PERRY, D.A.: Effects of increased rainfall on native forage production in eastern Montana. - J. Range Manage. 33: 246-250, 1980.

☆7534 - NEWBERY, D.M.: The effects of decomposing roots on the growth of grassland plants.- J. appl. Ecol. 16: 613-622, 1979.

7535 - NEWVILLE, E.G., FERRELL, W.K.: Abscisic acid levels and stomatal behavior during drought and recovery in Douglas-fir (Pseudotsuga menziesii). - Can. J. Bot. 58: 1370-1375, 1980.

7536 - NG, E., MILLER, P.C.: Soil moisture relations in the southern California chaparral. - Ecology 61: 98-107, 1980.

7537 - NIELSEN, D.C., SHAW, R.H.: Irrigation potential on Iowa soils of high water--holding capacity. - Iowa State J. Res. 54: 329-338, 1980.

7538 - NIKOLAEV, G.M., KNOX, P.P., KONONENKO, A.A., GRISHANOVA, N.P., RUBIN, A.B.: Photo-induced electron transport and water state in Rhodospirillum rubrum chromatophores. - Biochim.biophys. Acta 590: 194-201, 1980.

☆7539 - NIKOLSKI, Y.N.: The dependence of irrigation requirements on water-table depth in drained lands. - Agr. Water Manage. 1: 191-196, 1977.

☆7540 - NILSEN, K.N.: Enhanced stomatal conductance, water consumption, and growth of tomato (Lycopersicon esculentum Mill.) in response to far-red irradiance supplementation within controlled environments. - Plant Physiol. 63 (Suppl.): 126, 1979.

☆7541 - NIMBALKAR, J.D., JOSHI, G.V.: Effect of salt stress on photosynthesis in sugarcane var. Co. 740. - Biovigyanam 2: 137-144, 1976.

☆7542 - NIQUEUX, M.: Production d'été de variétés de Fétuque élevée en fonction des facteurs climatiques. - Ann. Amélior. Plant. 29: 97-105, 1979.

☆7543 - NIXON, P.R., SMITHEY, R.E.: Potential evapotranspiration in the Lower Rio Grande Valley. - J. Rio Grande Valley hort. Soc. 33: 91-100, 1979.

7544 - NOBEL, P.S.: Leaf anatomy and water use efficiency. - In: TURNER, N.C., KRAMER, P.J. (ed.): Adaptation of Plants to Water and High Temperature Stress. Pp. 43-55. John Wiley & Sons, New York - Chichester - Brisbane - Toronto 1980.

7545 - NOBEL, P.S.: Water vapor conductance and CO_2 uptake for leaves of a C_4 desert grass, Hilaria rigida. - Ecology 61: 252-258, 1980.

7546 - NOBEL, P.S.: Interception of photosynthetically active radiation by cacti of different morphology. - Oecologia 45: 160-166, 1980.

☆7547 - NOKS, P.P., KONONENKO, A.A., RUBIN, A.B.: Funktsional'naya aktivnost' fotosinteticheskikh reaktsionnykh tsentrov iz Rhodopseudomonas sphaeroides pri fiksirovannoĭ gidratatsii preparatov.[Function activity in photosynthetic reaction centers from Rhodopseudomonas sphaeroides at fixed hydration levels of the preparations.] - Bioorg. Khim. 5: 879-885, 1979. [In R, ab: E.]

7548 - NOKS, P.P., KONONENKO, A.A., RUBIN, A.B.: Vliyanie deĭterirovaniya na kinetiku fotoindutsirovannogo perenosa ėlektrona v reaktsionnykh tsentrakh purpurnykh bakteriĭ. [Effect of deuteration on the kinetics of photoinduced electron transfer in reaction centres of purple bacteria.] - Biofizika 25: 239-241, 1980. [In R, ab: E.]

☆7549 - NOLAND, T.L., KOZLOWSKI, T.T.: Effect of SO_2 on stomatal aperture and sulfur uptake of woody angiosperm seedlings. - Can. J. Forest Res. 9: 57-62, 1979.

*7550 - NOLAND, T.L., KOZLOWSKI, T.T.: Influence of potassium nutrition on suscepti-
bility of silver maple to ozone. - Can. J. Forest Res. 9: 501-503, 1979.

7551 - NORLYN, J.D.: Breeding salt-tolerant crop plants. - In: RAINS, D.W., VALENTINE,
R.C., HOLLAENDER, A. (ed.): Genetic Engineering of Osmoregulation. Impact on
Plant Productivity for Food, Chemicals, and Energy. Pp. 293-309. Plenum Press,
New York - London 1980.

7552 - NOWAKOWSKI, W.: Hipotezy dotyczące mechanizmu działania auksyn z uwzględnie-
niem gospodarki wodnej róslin. [Hypotheses concerning the mechanism of effect
of auxine taking into account plant water regime.] - Wiadomości bot. 24: 105-
112, 1980. [In Pol.]

7553 - NUKAYA, A., MASUI, M., ISHIDA, A,: Salt tolerançe of muskmelons in sand and
nutrient solution cultures. - J. Jap. Soc. hort. Sci. 49: 93-101, 1980.

7554 - NULSEN, R.A., THURTELL, G.W.: Effects of osmotica around the roots on water
uptake by maize plants. - Aust. J. Plant Physiol. 7: 27-34, 1980.

*7555 - OBYDENNYĬ, P.T.: Svyaz' intensivnosti fotosinteza drevesnykh rasteniĭ s fakto-
rami sredy v usloviyakh promyshlennogo zagryazneniya atmosfery. [Relation of
photosynthetic rate of woody plants with environmental factors under industri-
al pollution of atmosphere.] - In: NASYROV, Yu.S. (ed.): Genetika Fotosinteza.
Pp. 271-277. Donish, Dushanbe 1977. [In R.]

*7556 - OECHEL, W.C., LAWRENCE, W.T.: Energy utilization and carbon metabolism in
mediterranean scrub vegetation of Chile and California. I. Methods: A trans-
portable cuvette field photosynthesis and data acquisition system and representative results for *Ceanothus greggii*. - Oecologia 39: 321-335, 1979.

*7557 - OGAWA, T.: Two steps of gas exchange in leaf photosynthesis. - Physiol. Plant.
35: 91-95, 1975.

7558 - OGAWA, T.: Synergistic action of red and blue light on stomatal opening of
Vicia faba leaves. - In: SENGER, H. (ed.): The Blue Light Syndrome. Pp. 622-
628. Springer-Verlag, Berlin - Heidelberg - New York 1980.

7559 - OGAWA, T.: Effect of anaerobiosis on photosynthesis of higher plants. - Photo-
biochem. Photobiophys. 1: 321-328, 1980.

*7560 - OGAWA, T., ISHIKAWA, H., SHIMADA, K., SHIBATA, K.: Synergistic action of red
and blue light and action spectra for malate formation in guard cells of *Vicia
faba* L. - Planta 142: 61-65, 1978.

*7561 - OGAWA, T., SHIBATA, K.: Two phases of CO_2 absorption on leaves. - In: MITSUI,
A., MIYACHI, S., SAN PIETRO, A., TAMURA, S. (ed.): Biological Solar Energy
Conversion. Pp. 183-195. Academic Press, New York - San Francisco - London
1977.

*7562 - OHTOMO, K., FUJIMOTO, N., MINO, Y.: [Effect of soil compaction on the early
growth of roots of herbage plants.] - Res. Bull. Obihiro Univ. 11: 107-112,
1978. [In Jap, ab: E.]

*7563 - OHTOMO, K., TOMIYA, N., MINO, Y., SAKAI, R.: Effect of soil compaction on the
early growth of herbage plants. - Res. Bull. Obihiro Univ. 11: 491-497, 1979.

7564 - OKANENKO, A.A., KOMARENKO, N.I., TARAN, N.Yu.: Deĭstvie povyshennoĭ temperatury
v usloviyakh raznoĭ uvlazhnennosti pochvy na sostav lipidov v prorostkakh
pshenitsy. [Effect of elevated temperature under conditions of different mois-
tening of soil on composition of lipids in wheat seedlings.] - Fiziol. Biokhim. kul't. Rast. 12: 120-124, 1980. [In R, ab: E.]

*7565 - OKE, T.R.: Advectively-assisted evapotranspiration from irrigated urban vege-
tation. - Boundary-Layer Meteorol. 17: 167-173, 1979.

7566 - OLADOKUN, M.A.O.: Legume cover crops, organic mulch and associated soil conditions, and plant nutrient content for establishing quillou coffee. - Hort-
Science 15: 305-306, 1980.

✲7567 - OLECH, K.: Influence of the absence of light on the ensuing photosynthetic
activity of the leaves of some higher plants. - Pol. ecol. Stud. 1: 65-70,
1975.

✲7568 - OLLERENSHAW, J.H., HODGSON, D.R.: The effects of constant and varying heights
of cut on the yield of Italian ryegrass (*Lolium multiflorum* Lam.) and perennial ryegrass (*Lolium perenne* L.) - J. agr. Sci. 89: 425-435, 1977.

7569 - ONG, H.T.: Effects of actinomycin D, cycloheximide and kinetin on ribonuclease
and beta-fructofuranosidase in water stressed tomato cotyledons. - Biol. Plant.
22: 245-248, 1980.

7570 - ONG, H.T.: Effects of manitol induced water stress on the ribosomes of intact
leaves of Azuki·bean seedlings. - Biol. Plant. 22: 249-254, 1980.

✲7571 - OORSCHOT, J.L.P., van, LEEUWEN, P.H., van: Recovery from inhibition of photosynthesis by metamitron in various plant species. -·Weed Res. 19: 63-67, 1979.

7572 - ÖQUIST, G., BRUNES, L., HÄLLGREN, J.-E., GEZELIUS, K., HALLÉN, M., MALMBERG,
G.: Effects of artificial frost hardening and winter stress on net photosynthesis, photosynthetic electron transport and RuBP carboxylase activity in
seedlings of *Pinus silvestris*. - Physiol. Plant. 48: 526-531, 1980.

7573 - ÖQUIST, G., FORK, D.C.: Effects of desiccation on the excitation energy distribution between the two photosystems in the red alga *Porphyra perforata*. -
Carnegie Institute Washington Year Book 79: 193-197, 1980.

✲7574 - ORTON, P.J.: The influence of water stress and abscisic acid on the root development of *Chrysanthemum morifolium* cuttings during propagation. - J. hort.
Sci. 54: 171-180, 1979.

✲7575 - ORWICK, P.L., SCHREIBER, M.M.: Interference of redroot pigweed (*Amaranthus
retroflexus*) and robust foxtail (*Setaria viridis* var. *robusta-alba* or var.
robusta-purpurea) in soybeans (*Glycine max*). - Weed Sci. 27: 665-674, 1979.

7576 - OSMOND, C.B.: Integration of photosynthetic carbon metabolism during stress.
- In: RAINS, D.W., VALENTINE, R.C., HOLLAENDER, A. (ed.): Genetic Engineering
of Osmoregulation. Impact of Plant Productivity for Food, Chemicals, and Energy. Pp. 171-185. Plenum Press, New York - London 1980.

7577 - OSMOND, C.B., WINTER, K., POWLES, S.B.: Adaptive significance of carbon diox-
ide cycling during photosynthesis in water-stressed plants. - In: TURNER, N.
C., KRAMER, P.J. (ed.): Adaptation of Plants to Water and High Temperature
Stress. Pp. 139-154. John Wiley & Sons, New York - Chichester - Brisbane -
Toronto 1980.

7578 - OSONUBI, O., DAVIES, W.J.: The influence of water stress on the photosynthetic
performance and stomatal behaviour of tree seedlings subjected to variation
in temperature and irradiance. - Oecologia 45: 3-10, 1980.

7579 - OSONUBI, O., DAVIES, W.J.: The influence of plant water stress on stomatal
control of gas exchange at different levels of atmospheric humidity. - Oecologia 46: 1-6, 1980.

7580 - O'SULLIVAN, J.: Irrigation, spacing and nitrogen effects on yield and quality
of pickling cucumbers grown for mechanical harvesting. - Can. J. Plant Sci.
60: 923-928, 1980.

7581 - OTHIENO, C.O.: Effects of mulches on soil water content and water status of
tea plants in Kenya. - Exp. Agr. 16: 295-302, 1980.

7582 - O'TOOLE, J.C., CRUZ, R.T.: Response of leaf water potential, stomatal resistance, and leaf rolling to water stress. - Plant Physiol. 65: 428-432, 1980.

*7583 - PÁLFI, G., NÉMETH, J., PINTÉR, L., KÁDÁR, K., BÖLKE, W.: Rapid determination of drought-resistance of new rye, maize and lupine varieties with the live-
-wilting proline test. - Acta biol. (Szeged) 24: 39-51, 1978.

7584 - PALIWAL, K.V., YADAV, B.R.: Effect of saline irrigation water on the yield of potato. - Ind. J. agr. Sci. 50: 31-33, 1980.

*7585 - PALLARDY, S.G., KOZLOWSKI, T.T.: Frequency and length of stomata of 21 *Populus* clones. - Can. J. Bot. 57: 2519-2523, 1979.

7586 - PALLARDY, S.G., KOZLOWSKI, T.T.: Cuticle development in the stomatal region of *Populus* clones. - New Phytol. 85: 363-368, 1980.

7587 - PALLAS, J.E.,Jr.: An apparent anomaly in peanut leaf conductance. - Plant Physiol. 65: 848-851, 1980.

*7588 - PALLAS, J.E.,Jr., MICHEL, B.E.: Comparison of leaf and stem hygrometers for measuring changes in peanut plant water potential. - Peanut Sci. 5: 65-67, 1978.

7589 - PALMER, C.J., BLANCHAR, R.W.: Modification of Tempe pressure cell for the measurement of saturated hydraulic conductivities. - Soil Sci. Soc. Amer. J. 44: 430-431, 1980.

*7590 - PALTA, J.P., LI, P.H.: Frost-hardiness in relation to leaf anatomy and natural distribution of several *Solanum* species. - Crop Sci. 19: 665-671, 1979.

7591 - PALTA, J.P., LI, P.H.: Alterations in membrane transport properties by freezing injury in herbaceous plants: Evidence against rupture theory. - Physiol. Plant. 50: 169-175, 1980.

7592 - PALTA, J.P., STADELMANN, E.J.: On simultaneous transport of water and solute through plant cell membranes: Evidence for the absence of solvent drag effect and insensitivity of the reflection coefficient. - Physiol. Plant. 50: 83-90, 1980.

7593 - PĂLTINEANU, R., PĂLTINEANU, I.: Evapotranspiraţia determinată în lizimetre la unele plante furajere perene. [Evapotranspiration determined in lysimeters at some perennial fodder plants.] - An. Inst. Cercetări Cereale Plante tehnice Fundulea 45: 349-359, 1980. [In Roum. ab: R,E.]

7594 - PAQUIN, R., MEHUYS, G.R.: Influence of soil moisture on cold tolerance of alfalfa. - Can. J. Plant Sci. 60: 139-147, 1980.

7595 - PARKINSON, K.J., DAY, W.: Temperature corrections to measurements made with continuous flow porometers. - J. appl. Ecol. 17: 457-460, 1980.

7596 - PARKINSON, K.J., DAY, W., LEACH, J.E.: A portable system for measuring the photosynthesis and transpiration of graminaceous leaves. - J. exp. Bot. 31: 1441-1453. 1980.

7597 - PAROSCHY, J.H., MEIERING, A.G., PETERSON, R.L., HOSTETTER, G., NEFF, A.: Mechanical winter injury in grapevine trunks. - Amer. J. Enol. Viticult. 31: 227-232, 1980.

7598 - PARRA, M.A., CRUZ ROMERO, G.: On the dependence of salt tolerance of beans (*Phaseolus vulgaris* L.) on soil water potentials. - Plant Soil 56: 3-16, 1980.

*7599 - PARSONS, L.R., LI, P.H.: Changes in frost hardiness of stem cortical tissues of *Cornus stolonifera* Michx. after recovery from water stress. - Plant Physiol. 64: 351-353, 1979.

7600 - PASSIOURA, J.B.: The transport of water from soil to shoot in wheat seedlings. - J. exp. Bot. 31: 333-345, 1980.

7601 - PASSIOURA, J.B.: The meaning of matric potential. - J. exp. Bot. 31: 1161-1169, 1980.

*7602 - PATAKY, S.M., HORVÁTH, I.: The effect of covering with a transparent plastic sheet on the tissue structure of the leaves of bean plants. - Acta biol. (Szeged) 24: 31-38, 1978.

7603 - PATE, J.S., LAYZELL, D.B., ATKINS, C.A.: Transport exchange of carbon, nitrogen and water in the context of whole plant growth and functioning - case history of a nodulated annual legume. - Ber. Deut. bot. Ges. 93: 243-255, 1980.

7604 - PATE, J.S., LAYZELL, D.B., McNEIL, D.L.: Modeling the transport and utilization of carbon and nitrogen in a nodulated legume. - Plant Physiol. 63: 730-737, 1979.

7605 - PATEL, C.S.: Estimation of evapotranspiration of crops by indirect methods in multiple cropping of jute-rice-wheat. - Oryza 16: 32-36, 1979.

*7606 - PATEL, M.S., SINGH, N.T.: The effect of soil compaction on growth and water use efficiency of rice. - Ind. J. Agron. 24: 429-431, 1979.

7607 - PATEL, P.M., WALLACE, A., HARTSOCK, T., ROMNEY, E.M.: Zinc, nickel, and cadmium uptake and translocation to seed pods and their effects on gas exchange rates of bush bean plants grown in calcareous soil from the northern Mojave desert. - J. Plant Nutr. 2: 67-72, 1980.

*7608 - PATERSON, D.R., EARHART, D.R. , FUQUA, M.C.: Effects of flooding level on storage root formation, ethylene production, and growth of sweet potato. - HortScience 14: 739-740, 1979.

7609 - PATON, D.M., DHAWAN, A.K., WILLING, R.R.: Effect of *Eucalyptus* growth regulators on the water loss from plant leaves. - Plant Physiol. 66: 254-256, 1980.

*7610 - PATTERSON, R.P., RAPER, C.D.,Jr., GROSS, H.D.: Growth and specific nodule activity of soybean during application and recovery of a leaf moisture stress. - Plant Physiol. 64: 551-556, 1979.

7611 - PAVANASASIVAM, V., AXLEY, J.H.: Influence of flooding on the availability of soil zinc. - Commun. Soil Sci. Plant Anal. 11: 163-174, 1980.

7612 - PAVLIK, B.M.: Patterns of water potential and photosynthesis of desert sand dune plants, Eureka Valley, California. - Oecologia 46: 147-154, 1980.

*7613 - PEARCE, A.J., ROWE, L.K.: Forest management effects on interception, evaporation, and water yield. - J. Hydrol. (N. Zeal.) V18 (2): 73-87, 1979.

7614 - PEARCE, A.J., ROWE, L.K., STEWART, J.B.: Nighttime, wet canopy evaporation rates and the water balance of an evergreen mixed forest. - Water Resour. Res. 16: 955-959, 1980.

*7615 - PECK, R.A., KIRKHAM, M.B.: Water relations and yield of winter wheat grown under three water regimes in the High Plains. - Proc. Oklahoma Acad. Sci. 59: 53-59, 1979.

7616 - PEET, M.M., KRAMER, P.J.: Effects of decreasing source/sink ratio in soybeans on photosynthesis, photorespiration, transpiration and yield. - Plant Cell Environ. 3: 201-206, 1980.

7617 - PEISKER, M., VÁCLAVÍK, J.: Relationship between transpiration and CO_2 uptake in leaves of *Zea mays* L. after excision. - Photosynthetica 14: 545-549, 1980.

7618 - PELEVINA, L.V.: Vliyanie margantsa i tsinka na vodouderzhivayushchuyu sposob-
       nost' list'ev i urozhaĭnost' yabloni. [Effect of manganese and zinc on water-
       -retaining capacity of apple-tree leaves and yield.] - Fiziol. Biokhim. kul't.
       Rast. 12: 319-322, 1980. [In R, ab: E.]

*7619 - PEMADASA, M.A.: Stomatal movements in cotyledons. - Ceylon J. Sci., Biol. Sci.
       12: 1-8, 1976.

*7620 - PENKA, M., ČERMÁK, J., PALÁT, M.: Behaviour of the transpiration flow rate and
       its variations due to weather conditions observed in a full-grown tree *Prunus
       avium* L. - Acta Univ. agr. (Brno), Ser. C 45: 123-147, 1976.

*7621 - PEREIRA, J.S.: Relações hídricas das árvores. [Water relations of trees.] -
       Agron. lusit. 39: 155-173, 1979. [In Port, ab: E.]

*7622 - PERL, M.: Invigoration of cotton seedlings by treatment of seeds for pregermi-
       nation activities. - J. exp. Bot. 30: 183-192, 1979.

7623 - PERRY, K.B., MARTSOLF, J.D., MORROW, C.T.: Conserving water in sprinkling for
       frost protection by intermittent application. - J. Amer. Soc. hort. Sci. 105:
       657-660, 1980.

7624 - PERRY, L.P., BOODLEY, J.W.: Germination of foliage plant seeds in response to
       pre-sowing ultrasonic exposures, water soaks and fungicides. - HortScience 15:
       192-194, 1980.

7625 - PETERSON, J.C., SACALIS, J.N., DURKIN, D.J.: Alterations in abscisic acid con-
       tent of *Ficus benjamina* leaves resulting from exposure to water stress and its
       relationship to leaf abscision. - J. Amer. Soc. hort. Sci. 105: 793-798, 1980.

7626 - PETERSON, J.C., SACALIS, J.N., DURKIN, D.J.: Promotion of leaf abscision in
       intact *Ficus benjamina* by exposure to water stress. - J. Amer. Soc. hort. Sci.
       105: 788-793, 1980.

*7627 - PETERSON, J.R., COOPER, P.G.: Some considerations of water in the germination
       test. - Seed Sci. Technol. 7: 329-340, 1979.

7628 - PETOLINO, J.F., LEONE, I.A.: Saline aerosol: Some effects on the physiology of
       Phaseolus vulgaris. - Phytopathology 70: 229-232, 1980.

7629 - PEUKERT, D.E.: Zur Anatomie von *Epiphyllum chrysocardium* Alexander (*Cactaceae*)
       - Epidermis und Stomatogenese. - Flora 169: 1-8, 1980.

7630 - PEVELING, E., ROBENEK, H.: The plasmalemma structure in the phycobiont *Trebou-
       xia* at different stages of humidity of a lichen thallus. - New Phytol. 84:
       371-374, 1980.

7631 - PHAM THI, A.T., VIEIRA DA SILVA, J.: Influence de la sécherésse sur l'ultra-
       structure mitochondriale chez le cotonnier - Quelques implications métaboli-
       ques. - Z. Pflanzenphysiol. 100: 351-358, 1980.

*7632 - PHENE, C.J., BEALE, O.W.: Influence of twin-row spacing and nitrogen rates
       on high-frequency trickle-irrigated sweet corn. - Soil Sci. Soc. Amer. J. 43:
       1216-1221, 1979.

7633 - PHILPOTTS, H.R.: Effects of initial watering on nodulation of lupins. - Aust.
       J. exp. Agr. anim. Husb. 20: 599-601, 1980.

7634 - PICARD, D., COUCHAT, P., MOUTONNET, P.: Données préliminaires sur la transpi-
       ration du riz pluvial, variété IRAT 13, soumis a'une carence hydrique. - Plant
       Soil 57: 423-430, 1980.

*7635 - PICKARD, W.F., MINCHIN, P.E.H., TROUGHTON, J.H.: Real time studies of carbon-11
translocation in moonflower. III. Further experiments on the effects of a nitrogen atmosphere, water stress, and chilling; and a qualitative theory of
stem translocation. - J. exp. Bot. 30: 307-318, 1979.

*7636 - PIECHA, W.: Przebieg fotosyntezy i transpiracji u niektórych róslin motylko-
wych. [Course of photosynthesis and transpiration of some legumes.] - Zesz.
nauk. Akad. rol.-tech. Olsztynie, Rolnictwo 22: 11-18, 1977. [In Pol, ab: E.]

7637 - PIERCE, M., RASCHKE, K.: Correlation between loss of turgor and accumulation
of abscisic acid in detached leaves. - Planta 148: 174-182, 1980.

7638 - PILL, W.G., LAMBETH, V.N.: Effects of soil water regime and nitrogen form on
blossom-end rot, yield, water relations, and elemental composition of tomato.
- J. Amer. Soc. hort. Sci. 105: 730-734, 1980.

7639 - PIZZOLATO, T.D.: On the vascular anatomy and stomates of the lodicules of Zea
mays. - Can. J. Bot. 58: 1045-1055, 1980.

7640 - PLHÁK, F.: Změny v obsahu nestrukturních glycidů při fotosyntéze vojtěšky.
[Changes in content of nonstructural carbohydrates during photosynthesis of
alfalfa.] - In: Dny Rostlinné Fyziologie II. Pp. 310-313. Vysoká Škola Zemědělská, Brno 1980. [In Czech, ab: E,R.]

7641 - PÓCS, T.: The epiphytic biomass and its effect on the water balance of two rain
forest types in the Uluguru Mountains (Tanzania, East Africa). - Acta bot.
Acad. Sci. Hung. 26: 143-167, 1980.

*7642 - POLLARD, A., WYN JONES, R.G.: Enzyme activities in concentrated solutions of
glycinebetaine and other solutes. - Planta 144: 291-298, 1979.

*7643 - POLONSKIĬ, V.I., ZINENKO, G.K., GRIBOVSKAYA, I.V.: Dinamika vynosa mineral'-
nykh elementov pshenitseĭ v usloviyakh intensivnogo kul'tivirovaniya. [Dynamics
of use of mineral elements by wheat plants under conditions of intensive cultivation.] - In: Intensivnaya Svetokul'tura Rasteniĭ. Pp. 48-58. Inst. Fiz.
Sib. Otd. Akad. Nauk SSSR, Krasnoyarsk 1977. [In R.]

7644 - POSGAY, E.: A vízellátás és a termés közötti kapcsolat az öntözéses növényter-
mesztésben I. Szója. [Correlation between water supply and yield in crop
growing under irrigation I. Soybeans.] - Növénytermelés 29: 465-475, 1980.
[In Hung, ab: E.]

7645 - POSPÍŠILOVÁ, J., SOLÁROVÁ, J.: Environmental and biological control of diffusive conductances of adaxial and abaxial leaf epidermes. - Photosynthetica 14:
90-127, 1980.

7646 - POTTER, J.R., BREEN, P.J.: Maintenance of high photosynthetic rates during the
accumulation of high leaf starch levels in sunflower and soybean. - Plant
Physiol. 66: 528-531, 1980.

*7647 - PRAGA-KUBALSKA, B., GEJ, B.: Asymilacja $^{14}CO_2$ i przebieg wzrostu pszenicy ja-
rej odm. Nagradowicka w warunkach intensywnego zywienia azotem lub fosforem.
[Assimilation of $^{14}CO_2$ and growth dynamics of spring wheat cv. Nagradowicka
under conditions of intensive nitrogen or phosphorus fertilization.] - Acta
agrobot. 31: 107-115, 1978. [In Pol, ab: E.]

*7648 - PREMAKUMARI, D., ANNAMMA, Y., BHASKARAN NAIR, V.K.: Clonal variability for
stomatal characters and its application in Hevea breeding and selection. -
Ind. J. agr. Sci. 49: 411-413, 1979.

*7649 - PŘIBÁŇ, K., ONDOK, J.P.: The daily and seasonal course of evapotranspiration
from a central european sedge-grass marsh. - J. Ecol. 68: 547-559, 1980.

7650 - PŘIBÁŇ, K., ONDOK, J.P.: Sezónní měření radiační a tepelné bilance na "Mokrých lukách" u Třeboně. [Seasonal measuring of radiation and temperature balance on "Mokrá luka" ("Wet Meadows") near Třeboň.] - In: Bilancia Energie a Vody v Poľných a Lesných Ekosystémoch. Pp. 32-41. Vysoká Škola Poľnohospodárská, Nitra 1979. [In Czech]

*7651 - PRIEHRADNÝ, S.: Changes in water balance by powdery mildew infected susceptible barley cultivar. - Acta phytopathol. Acad. Sci. Hung. 14: 351-361, 1979.

*7652 - PRIEHRADNÝ, S.: Changes in water balance by powdery mildew infected resistant barley cultivar. - Acta phytopathol. Acad. Sci. Hung. 14: 363-368, 1979.

7653 - PRIEHRADNÝ, S.: Zmeny vzťahov príjmu vody a transpirácie po infekcii patogénom náchylného a rezistentného hostitela. [Changes in the relations of water uptake and transpiration after infection by a pathogen of susceptible and resistant hosts.] - Biológia (Bratislava) 35: 717-725, 1980. [In Slov, ab: E,R.]

7654 - PRIOUL, J.-L., BRANGEON, J., REYSS, A.: Interaction between external and internal conditions in the development of photosynthetic features in a grass leaf II. Reversibility of light-induced responses as a function of developmental stages. - Plant Physiol. 66: 770-774, 1980.

*7655 - PRITCHARD, D.T., Carbon dioxide production in soils, under barley, subjected to a range of drought treatments. - J. Sci. Food Agr. 30: 547-557, 1979.

*7656 - PROCTOR, M.C.F.: Surface wax on the leaves of some mosses. - J. Bryol. 10: 531-538, 1979.

7657 - PTÁČKOVÁ, M.: Transpirace vojtěšky v průběhu vegetačního období. [Lucerne transpiration during the growing season.] - Rost. Výroba (Praha) 26: 143-151, 1980. [In Czech, ab: E,G,R.]

7658 - PUARD, M., COUCHAT, P., MOUTONNET, P.: Application de la methode gamma neutronique a une etude d'infiltration d'eau sous riziere. - Agron. trop. 35: 25 -29, 1980.

7659 - PUROHIT, S.S.: Effects of Dikegulac-sodium on stomatal movement of *Helianthus annuus*. - Comp. Physiol. Ecol. 5: 159, 1980.

*7660 - PYYKKÖ, M.: Morphology and anatomy of leaves from some woody plants in a humid tropical forest of Venezuelan Guayana. - Acta bot. Fenn. 112: 1-41, 1979.

7661 - QUARRIE, S.A.: Genotypic differences in leaf water potential, abscisic acid and proline concentrations in spring wheat during drought stress. - Ann. Bot. 46: 383-394, 1980.

*7662 - QUEIROZ de VILHENA, R.C.: Anatomia foliar de três espécies da família *Humiriaceae*. [Leaf anatomy of tree species of *Humiriaceae*.] - Acta Amazonica 8: 25-43, 1978. [In Port, ab: E.]

7663 - QUISENBERRY, J.E., ROARK, B., FRYREAR, D.W., KOHEL, R.J.: Effectiveness of selection in upland cotton in stress environments. - Crop Sci. 20: 450-453, 1980.

7664 - RADCLIFFE, D., HAYDEN, T., WATSON, K., CROWLEY, P., PHILLIPS, R.E.: Simulation of soil water within the root zone of a corn crop. - Agron. J. 72: 19-24, 1980.

*7665 - RADEVA, V.G.: Water deficit in lucerne. III. Water deficit and N, P, K, Ca and Mg uptake by lucerne plants. - In: KUDREV, T., STOYANOV, I., GEORGIEVA, V. (ed.): Mineral Nutrition of Plants. Vol. II. Pp. 236-239. Publishing House Central Cooperative Union, Sofia 1979.

7666 - RAGHAVENDRA, A.S.: Chloride and nitrate stimulate stomatal opening and de-
crease potassium uptake and malate production in epidermal tissues of *Commelina benghalensis*. - Aust. J. Plant Physiol. 7: 663-669, 1980.

7667 - RAGUSE, C.A., HULL, J.L., DELMAS, R.E.: Perennial irrigated pastures. III.
Beef calf production from irrigated pasture and winter annual range. - Agron.
J. 72: 493-499, 1980.

7668 - RAHMAN, M.S., RUTTER, A.J.: A comparison of ecology of *Deschampsia cespitosa*
and *Dactylis glomerata* in relation to the water factor II. Controlled experiments in glasshouse conditions. - J. Ecol. 68: 479-491, 1980.

*7669 - RAI, R.S.V., MURTY, K.S.: Effect of submergence on some physiological changes
in rice seedlings. - Ind. J. exp. Biol. 14: 369-370, 1976.

*7670 - RAI, R.S.V., MURTY, K.S.: Note on the correlation of yield with yield attrib-
utes in rice under two water regimes and fertilizer rates. - Ind. J. agr. Sci.
49: 812-814, 1979.

*7671 - RAINS, D.W.: Salt tolerance of plants: Strategies of biological systems. -
In: HOLLAENDER, A., ALLER, J.C., EPSTEIN, E., SANPIETRO, A., ZABORSKY, O.R.
(ed.): Biosaline Concept. An Approach to the Utilization of Underexploited
Resources. Plenum Press, New York - London 1979.

7672 - RAINS, D.W., CROUGHAN, T.P., STAVAREK, S.J.: Selection of salt-tolerant plants
using tissue culture. - In: RAINS, D.W., VALENTINE, R.C., HOLLAENDER, A. (ed.):
Genetic Engineering of Osmoregulation. Impact on Plant Productivity for Food,
Chemicals, and Energy. Pp. 279-292. Plenum Press, New York - London 1980.

7673 - RAISON, J.K., BERRY, J.A., ARMOND, P.A., PIKE, C.S.: Membrane properties in
relation to the adaptation of plants to temperature stress. - In: TURNER, N.C.,
KRAMER, P.J. (ed.): Adaptation of Plants to Water and High Temperature Stress.
Pp. 261-273. John Wiley & Sons, New York - Chichester - Brisbane - Toronto
1980.

7674 - RAJ, S.A., SINGARAVADIVEL, K., VASAN, B.S., VENKATESAN, V.: Moisture reduction
in high-moisture rice. - Ind. J. agr. Sci. 50: 675-679, 1980.

7675 - RAJAGOPAL, V., ANDERSEN, A.S.: Water stress and root formation in pea cuttings.
I. Influence of the degree and duration of water stress on stock plants grown
under two levels of irradiance. - Physiol. Plant. 48: 144-149, 1980.

7676 - RAJAGOPAL, V., ANDERSEN, A.S.: Water stress and root formation in pea cuttings.
III. Changes in the endogenous level of abscisic acid and ethylene production
in the stock plants under two levels of irradiance. - Physiol. Plant. 48: 155-
160, 1980.

*7677 - RAKHMANINA, K.P., MOLOTKOVSKIĬ, Yu.I.: Vodnyĭ rezhim *Phragmites australis* v
Yuzhnom Tadzhikistane. [Water regime of *Phragmites australis* in South Taji-
kistan.] - Ékologiya 1979 (5): 22-32, 1979. [In R.]

7678 - RAMAGE, R.T.: Genetic methods to breed salt tolerance in plants. - In: RAINS,
D.W., VALENTINE, R.C., HOLLAENDER, A. (ed.): Genetic Engineering of Osmoregulation. Impact on Plant Productivity for Food, Chemicals, and Energy. Pp. 311
-318. Plenum Press, New York - London 1980.

7679 - RAMESH BABU, V.: Water uptake by tomato (*Lycopersicon esculentum*) leaves from
dew produced artificially. - Ind. J. Plant Physiol. 23: 10-13, 1980.

7680 - RAMI REDDY, S., HUKKERI, S.B.: Soil, water and weed management for direct-
-seeded rice grown on irrigation soils in north-western India. - Ind. J. agr.
Sci. 49: 427-433, 1979.

*7681 – RANA, B.S., MOHAN RAO, V.J., RAO, N.G.P.: Genetic analysis of some exotic × Indian crosses in sorghum XVIII. Breeding for resistance to grain deterioration. - Ind. J. Genet. Plant Breed. 38: 322-332, 1978.

7682 – RAND, R.H., COOKE, J.R.: A comprehensive model for CO_2 assimilation in leaves. - Trans ASAE 23: 601-607, 1980.

*7683 – RAO, I.M., SWAMY, P.M., DAS, V.S.R.: Some characteristics of Crassulacean Acid Metabolism in five nonsucculent scrub species under natural semiarid conditions. - Z. Pflanzenphysiol. 94: 201-210, 1979.

*7684 – RAO, I.M., SWAMY, P.M., DAS, V.S.R.: The reversal of scotoactive stomatal behavior in some woody weeds by paraquat and 2,4,5-T. - Weed Sci. 25: 469-472, 1977.

7685 – RAO, P.V.: Determination of a growth-environment relationship in jute (*Corchorus olitorius* L.) - Agr. Meteorol. 22: 45-52, 1980.

7686 – RAO, P.V.: Effects of rainfall and temperature on yield of tossa jute. - Ind. J. agr. Sci. 50: 608-611, 1980.

7687 – RASMUSSEN, S., ANDERSEN, A.S.: Water stress and root formation in pea cuttings. II. Effect of abscisic acid treatment of cuttings from stock plants grown under two levels of irradiance. - Physiol. Plant. 48: 150-154, 1980.

7688 – RAVEN, J.A., SMITH, F.A., SMITH, S.E.: Ions and osmoregulation. - In: RAINS, D. W., VALENTINE, R.C., HOLLAENDER, A. (ed.): Genetic Engineering of Osmoregulation. Impact on Plant Productivity for Food, Chemicals, and Energy. Pp. 101-118. Plenum Press, New York - London 1980.

7689 – RAWSON, H.M., CONSTABLE, G.A.: Carbon production of sunflower cultivars in field and controlled environments. I Photosynthesis and transpiration of leaves, stems and heads. - Aust. J. Plant Physiol. 7: 555-573, 1980.

7690 – RAWSON, H.M., CONSTABLE, G.A., HOWE, G.N.: Carbon production of sunflower cultivars in field and controlled environments. II Leaf growth. - Aust. J. Plant Physiol. 7: 575-586, 1980.

7691 – REAVES, R.E., KIRKHAM, M.B., TAYLOR, A.G., CAMPBELL, R.E.: Growth of cucumber under water and temperature stress. - J. Arid Environ. 3: 113-115, 1980.

*7692 – REDDY, P.K.R., SHAH, G.L.: Observations on the cotyledonary and hypocotyledonary stomata and trichomes in some *Caesalpiniaceae* with a note on their taxonomic significance. - Feddes Repert. 90: 239-250, 1979.

*7693 – REED, K.L., HAMERLY, E.R., DINGER, B.E., JARVIS, P.G.: An analytical model for field measurement of photosynthesis. - J. appl. Ecol. 13: 925-942, 1976.

7694 – REED, R.H., COLLINS, J.C., RUSSELL, G.: The influence of variations in salinity upon photosynthesis in the marine alga *Porphyra purpura* (Roth) C. Ag. (*Rhodophyta, Bangiales*). - Z. Pflanzenphysiol. 98: 183-187, 1980.

7695 – REICH, P.B., HINCKLEY, T.M.: Water relations, soil fertility, and plant nutrient composition of a pygmy oak ecosystem. - Ecology 61: 400-416, 1980.

7696 – REICOSKY, D.C., DEATON, D.E., PARSONS, J.E.: Canopy air temperatures and evapotranspiration from irrigated and stressed soybeans. - Agr. Meteorol. 21: 21-35, 1980.

7697 – REINERT, J. (ed.): Chloroplasts. (Results and Problems in Cell Differentiation. Vol. 10.) - Springer-Verlag, Berlin - Heidelberg - New York 1980.

7698 - REINHARDT, E., BURGER, G., WEISE, G.: Zur Erkundung der phytotoxischen Wirkung von Zink, Kadmium und Kupfer auf *Elodea canadensis* Michx. durch Erfassen der Deplasmolysezeit. - Acta hydrochim. hydrobiol. 8: 149-160, 1980.

7699 - REINHARDT, E., WEISE, G., BURGER, G.: Zellphysiologische Untersuchungen mittels *Elodea canadensis* Michx zur Biotropie diphosphonsäurehaltiger Stabilisatoren der Wasserhärte. - Acta Hydrochim. Hydrobiol. 8: 143-147, 1980.

*7700 - RESZEL, R.: Wpływ deszczowania i nawożenia fosforowo-potasowego na lucernę mieszańcową uprawiana na rędzinie. [Effect of sprinkler irrigation and phosphorus-potassium fertilization on hybrid alfalfa cultivated on rendzina soil.] - Rocz. Nauk roln., Ser. A 104: 91-106, 1979. [In Pol, ab: E,R.]

*7701 - REYNIERS, F.N., JACQUOT, M.: Demarche pour l'obtention de la resistance varietale a la secheresse. Cas du riz pluvial. - Agron. trop. 33: 314-317, 1978.

*7702 - RICHARDS, R.A., THURLING, N.: Genetic analysis of drought stress response in rapeseed (*Brassica campestris* and *B. napus*). II. Yield improvement and the application of selection indices. - Euphytica 28: 169-177, 1979.

*7703 - RICHARDS, R.A., THURLING, N.: Genetic analysis of drought stress response in rapeseed (*Brassica campestris* and *B. napus*). III. Physiological characters. - Euphytica 28: 755-759, 1979.

7704 - RICHARDSON, S.G., McKELL, C.M.: Water relations of *Atriplex canescens* as affected by the salinity and moisture percentage of processed oil shale. - Agron. J. 72: 946-950, 1980.

*7705 - RICHTER, W.: Mehrjährige Ergebnisse zum Einfluss der Beregnung auf die Ertragsleistungen von Getreidearten auf einem grundwasserfernen leichten Diluvial-standort. - Arch. Acker- Pflanzenbau Bodenk. 23: 617-624, 1979.

7706 - RICKMAN, R.W., KLEPPER, B.L.: Wet season aeration problems beneath surface mulches in dryland winter wheat production. - Agron. J. 72: 733-736, 1980.

7707 - RIGGLE, F.R., SLACK, D.C.: Rapid determination of soil water characteristic by thermocouple psychrometry. - Trans. ASAE 23: 99-103, 1980.

7708 - RIOUX, R., COMEAU, J.E.: Influence des systèmes de culture sur la croissance et le rendement des pommes de terre. - Can. J. Plant Sci. 60: 591-598, 1980.

7709 - RITCHIE, J.T.: Plant stress research and crop production: The challenge ahead. - In: TURNER, N.C., KRAMER, P.J. (ed.): Adaptation of Plants to Water and High Temperature Stress. Pp. 21-29, John Wiley & Sons, New York - Chichester - Brisbane - Toronto 1980.

7710 - RITCHIE, J.T.: Interaction and integration of adaptations to stress. - In: TURNER, N.C., KRAMER, P.J. (ed.): Adaptation of Plants to Water and High Temperature Stress. Pp. 447-450. John Wiley & Sons, New York - Chichester - Brisbane - Toronto 1980.

7711 - ROBERTS, J., PYMAR, C.F., WALLACE, J.S., PITMAN, R.M.: Seasonal changes in leaf area, stomatal and canopy conductances and transpiration from bracken below a forest canopy. - J. appl. Ecol. 17: 409-422, 1980.

7712 - ROBERTS, S.W., STRAIN, B.R., KNOERR, K.R.: Seasonal patterns of leaf water relations in four co-occuring forest tree species: parameters from pressure-volume curves. - Oecologia 46: 330-337, 1980.

7713 - ROBERTSON, W.K., HAMMOND, L.C., JOHNSON, J.T., BOOTE, K.J.: Effects of plant-water stress on root distribution of corn, soybeans, and peanuts in sandy soil. - Agron. J. 72: 548-550, 1980.

7714 - ROBINSON, D.G., QUADER, H.: Topographical features of the membrane of *Poterio-ochromonas malhamensis* after colchicine and osmotic treatment. - Planta 148: 84-88, 1980.

7715 - ROBINSON, F.E.: Irrigation rates critical in Imperial Valley alfalfa - Calif. Agr. 34 (10) : 18, 1980.

7716 - ROBINSON, J.T., HAMILTON, D.F.: Effects of time and rate of nutrient application on foliar nutrient concentration and cold hardiness in *Viburnum* species.- Scientia Hort. 13: 271-281, 1980.

7717 - ROBINSON, J.T., HAMILTON, D.F.: Effect of irradiance on nutrient uptake, growth and cold hardiness of *Viburnum opulus* L. 'Nanum'. - Scientia Hort. 13: 391-397, 1980.

*7718 - ROBINSON, S.J., YOCUM, C.F.: Photosynthetic properties of spheroplast preparations of the cyanobacterium *Spirulina platensis*. - Plant Physiol. 63 (Suppl.): 29, 1979.

7719 - RODE, J.C., BETHENOD, O.: Influence de la carence hydrique sur le comportement photosynthétique du Lin. - Ann. agron. 31: 285-295, 1980.

7720 - RODSKJER, N., SANDSBORG, J.: An experimental investigation of the energy balance for barley and winter wheat during the vegetation period. - Swed. J. agr. Res. 10: 155-158, 1980.

7721 - ROGERS, C., SHARPE, P.J.H., POWELL, R.D.: Dark opening of stomates of *Vicia faba* in CO_2-free air. Effect of temperature on stomatal aperture and potassium accumulation. - Plant Physiol. 65: 1036-1038, 1980.

7722 - ROLAND, J.-C., ROLAND, F.: Atlas of Flowering Plant Structure. - Longman, London - New York 1980.

*7723 - ROOK, D.A., SWANSON, R.H., CRANSWICK, A.M.: Reaction of radiata pine to drought. - N. Zeal. Dep. Sci. Indust. Res. Inform. Ser. 126 (Proceedings of Soil and Plant Water Symposium): 55-68, 1977.

*7724 - ROOS, E.E.: Germination of pelled and taped carrot and onion seed following storage. - J. Seed Technol. 4: 65-78, 1979.

*7725 - ROOS, E.E.: Storage behavior of pelled, tableted, and taped lettuce seed. - J. Amer. Soc. hort. Sci. 104: 283-288, 1979.

7726 - ROOS, E.E.: Physiological, biochemical, and genetic changes in seed quality during storage. - HortScience 15: 781-784, 1980.

*7727 - ROOS, E.E., JACKSON, G.S.: Testing coated seed: germination and moisture absorption properties. - J. Seed Technol. 1: 86-95, 1976.

*7728 - ROOS, E.E., SOWA, S., BURTON, G.W.: Accelerated aging studies of normal and segregating chlorophyl deficient isolines of pearl millet. - Crop Sci. 18: 231-233, 1978.

*7729 - ROSA, R.N.: A Energia Solar. Seu Aproveitamento por Conversao Fotossintética. [Solar Energy. Its Use for Photosynthetic Conversion.] - Lab. Ffs. Engenh. Nucl., LFEN-E-N°20, U.C.N., Sacavém, Portugal 1977. [In Port, ab: E.]

7730 - ROSS, H.A., HEGARTY, T.W.: Action of growth regulators on lucerne germination and growth under water stress. - New Phytol. 85: 495-501, 1980.

7731 - ROTH, D., KLEINSTÄUBER, G., WÖSS, W.: Untersuchungen zur Eignung elektrischer Bodenfeuchtemessverfahren. - Arch. Acker- Pflanzenbau Bodenk. 24: 411-416, 1980.

7732 - ROUQUETTE, F.M.,Jr., KEISLING, T.C., CAMP, B.J., SMITH, K.L.: Effect of drought stress on palatability of hybrid pearl millet. - Texas agr. Exp. Sta. Prog. Rep. 37733: 84-90, 1980.

7733 - ROY, J., METHY, M.: Absorption of photosynthetically active radiation by the leaves of grass species: inter- and intraspecific variations and phenotypic response to water stress. - Acta Oecol. - Oecol. Plant. 1: 253-256, 1980.

7734 - RUCKENBAUER, P., RICHTER, H.: Frictional resistances to water transport in water-cultured wheat plants. - Phyton 20: 37-45, 1980.

7735 - RUESS, R.W., WALI, M.K.: Daily fluctuations in water potential and associated ionic changes in *Atriplex canescens*. - Oecologia 47: 200-203, 1980.

☆7736 - RÜFFER-TURNER, M., NAPP-ZINN, K.: Investigations on leaf structure in several genotypes of *Arabidopsis thaliana* (L.) Heynh. - Arabidopsis Inform. Serv. 16: 94-98, 1979.

7737 - RUNDEL, P.W., EHLERINGER, J., MOONEY, H.A., GULMON, S.L.: Patterns of drought response in leaf-succulent shrubs of the coastal Atacama Desert in northern Chile. - Oecologia 46: 196-200, 1980.

7738 - RUNDEL, P.W., LANGE, O.L.: Water relations and photosynthetic response of a desert moss. - Flora 169: 329-335, 1980.

7739 - RUNNING, S.W.: Environmental and physiological control of water flux through *Pinus contorta*. - Can. J. Forest Res. 10: 82-91, 1980.

7740 - RUNNING, S.W.: Field estimates of root and xylem resistance in *Pinus contorta* using root excision. - J. exp. Bot. 31: 555-569, 1980.

7741 - RUNNING, S.W.: Relating plant capacitance to the water relations of *Pinus contorta*. - Forest Ecol. Manage. 2: 237-252, 1980.

7742 - RUNNING, S.W., REID, C.P.: Soil temperature influences on root resistance of *Pinus contorta* seedlings. - Plant Physiol. 65: 635-640, 1980.

7743 - RUSSELL, G.: Crop evaporation, surface resistance and soil water status. - Agr. Meteorol. 21: 213-226, 1980.

7744 - RYBALOVA, B.A., KATZ, K.M., PROKOF'EV, A.A.: Vzaimosvyaz' mezhdu transpiratsieǐ i soderzhaniem vody v sozrevayushchikh plodakh maka. [Relationship between transpiration and water content of ripening *Papaver somniferum* fruit.] - Fiziol. Rast. 27: 251-259, 1980. [In R, ab: E.]

7745 - RYBKINA, G.V., GUSEV, N.A., BIGLOVA, S.G.: Reaktsiya khloroplastov na obezvozhivanie. I. K fenomenu Brilliant. [Dehydration response of chloroplasts. I. I. On Brilliant phenomenon.] - Fiziol. Biokhim. kul't. Rast. 12: 291-297, 1980. [In R.]

7746 - RYČ, M., LEWAK, S.: The role of abscisic acid (ABA) in regulation of some photosynthetic enzyme activities in apple seedlings in relation to embryonal dormancy. - Z. Pflanzenphysiol. 96: 195-202, 1980.

7747 - RYCHNOVSKÁ, M., ČERMÁK, J., ŠMÍD, P.: Water output in a stand of *Phragmites communis* Trin. A comparison of three methods. - Acta Sci. Nat. Acad. Sci. Bohemoslov. (Brno) 14 (2): 1-30, 1980.

7748 - SAKAI, W.S., SANFORD, W.G.: Ultrastructure of the water-absorbing trichomes of pineapple (*Ananas comosus, Bromeliaceae*). - Ann. Bot. 46: 7-11, 1980.

7749 - SAKURATANI, T.: Apparent thermal conductivity of rice stems in relation to transpiration stream. - J. agr. Meteorol. 34: 177-187, 1979.

*7750 - **SALAMA, F.M.**: Photosynthesis and formation of the native chlorophyll forms in light harvesting complexes of wheat chloroplasts under physiological drought (salinity). - In: COOMBS, J. (ed.): 4th International Congress on Photosynthesis. P. 323a. UKISES, London 1977.

7751 - **SAMIEV, Kh.S., MARFINA, K.G.**: Kharakteristika belkov khloroplastov pri vodnom defitsite khlopchatnika. [Characteristics of chloroplast proteins in cotton plants grown under water stress.] - Fiziol. Rast. 27: 820-827, 1980. [In R, ab: E.]

7752 - **SAMMIS, T.W.**: Comparison of sprinkler, trickle, subsurface, and furrow irrigation methods for row crops. - Agron. J. 72: 701-704, 1980.

7753 - **SAMSUDDIN, Z.**: Differences in stomatal density, dimension and conductances to water vapour diffusion in seven *Hevea* species. - Biol. Plant. 22: 154-156, 1980.

*7754 - **SAMSUDDIN, Z., IMPENS, I.**: Relationship between photosynthetic rates and latex production in some *Hevea brasiliensis* Muel. Arg. cultivars. - In: COOMBS, J. (ed.): 4th International Congress on Photosynthesis. P. 323. UKISES, London 1977.

*7755 - **SAMUILOV, F.D.**: Vliyanie fosfornogo pitaniya na énergeticheskiĭ obmen i ustoĭchivost' rasteniĭ k neblagopriyatnym usloviyam sredy. [Effect of phosphorus nutrition on the energy metabolism and plant resistance to unfavorable environmental conditions.] - Izv. Akad. Nauk. SSSR, Ser. biol. 1978 (6): 828-839, 1978. [In R, ab: E.]

*7756 - **SÁNCHEZ CONDE, M.P., AZUARA, P.**: Variations in the composition of the sap of *Zea mays* with the increase in osmotic pressure of the nutritive solution. - J. Plant Nutr. 2: 305-322, 1979.

7757 - **SANCHEZ CONDE, M.P., AZUARA, P.**: Evaluacion del efecto de soluciones nutritivas equilibradas de distintas presiones osmoticas sobre la planta de lechuga. [Evaluation of the effect of osmotic pressure of nutritive solutions applied to a lettuce plant.] - Agrochimica 24: 176-183, 1980. [In Span, ab: E,F,G.]

7758 - **SANDS, R., REID, C.P.P.**: The osmotic potential of soil water in plant/soil systems. - Aust. J. Soil Res. 18: 13-25, 1980.

7759 - **SANDSBORG, J., OLOFSSON, B.**: Observations on the water balance for winter wheat, barley and fallow. - Swed. J. agr. Res. 10: 17-23, 1980.

7760 - **SANKHLA, R., CHAWAN, D.D.**: Effect of different seed moisture levels on the germination behaviour of *Phaseolus trilobus* Ait. - Biol. Plant. 22: 388-391, 1980.

7761 - **ŠANTA, M.**: Úloha zavlažovania pri intenzívnej výrobe krmovín v kukuričnej výrobnej oblasti. [The role of irrigation at intensive production of fodder crops in the maize-growing region.] - Rost. Výroba (Praha) 26: 1191-1196, 1980. [In: Slov, ab: E,R.]

*7762 - **SARICH, M., KASTORI, R., PETROVICH, M.**: Vliyanie kachestva sveta na pogloshchenie ^{32}P i ^{45}Ca molodymi rasteniyami kukuruzy. [Effect of light quality on the ^{32}P and ^{45}Ca absorption by young maize plants.] - Fiziol. Biokhim. kul't. Rast. 8: 411-414, 1976. [In R, ab: E.]

*7763 - **SASTRY, P.S.N., CHAKRAVARTY, N.V.K.**: Estimation of evapotranspiration for water balance studies in a semi-arid region. - Ind. J. Power River Valley Develop. 25: 288-291, 1975.

*7764 - **SAUGIER, B.**: Sunflower. - In: MONTEITH, J.L. (ed.): Vegetation and the Atmosphere. Vol. 2. Case Studies. Pp. 87-119. Academic Press, London - New York - San Francisco 1976.

*7765 - SAVOVA, N.P.: Efekt ot mineralnoto torene na lyutsernata pri razlichna stepen
na vodoobezpechenost. [Effect of mineral fertilization of lucerne at different
levels of water supply.] - Rasteniev. Nauki 12 (4): 88-95, 1975. [In Bulg, ab:
E,R.]

*7766 - SAVOVA, N.P., KAREV, K.S.: Izmenenie na produktivnostta na fotosintezata pri
pamuka pod vliyanie na napoyavaneto i toreneto. [Changes in the productivity
of photosynthesis in cotton under the influence of irrigation and fertilization.] - Rasteniev. Nauki 12 (2): 110-117, 1975. [In Bulg, ab: E,R.]

*7767 - SAXENA, M.C., LALORAYA, M.M.: Hormone induced water uptake in pea epicotyl
sections I: Effect of hormonal interactions. - Nat. Acad. Sci. Lett. 2 (11):
405-406, 1979.

*7768 - SAXENA, M.C., LALORAYA, M.M.: Hormone induced water uptake in pea epicotyl
sections. II: Effect of respiratory inhibitors. - Nat. Acad. Sci. Lett. 2 (12):
435-438, 1979.

7769 - SAXTON, M.J., BREIDENBACH, R.W., LYONS, J.M.: Membrane dynamics: Effects of
environmental stress. - In: RAINS, D.W., VALENTINE, R.C., HOLLAENDER, A. (ed.):
Genetic Engineering of Osmoregulation. Impact on Plant Productivity for Food,
Chemicals, and Energy. Pp. 203-233. Plenum Press, New York - London 1980.

*7770 - SCARASCIA MUGNOZZA, G.T., MONTI, L.M.: Plant breeding for yield under arid
conditions. - Genet. agr. 33: 331-340, 1979.

7771 - SCHÄFER, W.: Beziehungen zwischen CO_2-Assimilation und Wasserverbrauch. -
Arch. Acker- Pflanzenbau Bodenk. 24: 483-489, 1980.

7772 - SCHAFER, W., HOTZLER, I.: Photosynthese und aktuelle Evapotranspiration von
Winterweizen. - In: HOFFMANN, P., HIEKE, B. (ed.): Biophysik, Biochemie und
Physiologie der Photosynthese. Pp. 219-232. Humboldt-Universität, Berlin 1980.

7773 - SCHINDLER, U.: Ein Schnellverfahren zur Messung der Wasserleitfähigkeit im
teilgesättigten Boden an Stechzylinderproben. - Arch. Acker- Pflanzenbau
Bodenk. 24: 1-7, 1980.

7774 - SCHINDLER, U.: Die Kalibrierung der Neutronensonde für Wasserhaushaltsmessungen
in Lysimetern mit quellungsfähigem Bodenmaterial - ein Beitrag zur Methodik. -
Arch. Acker- Pflanzenbau Bodenk. 24: 553-559, 1980.

7775 - SCHLESINGER, W.H., GILL, D.S.: Biomass, production, and changes in the availability of light, water, and nutrients during the development of pure stands of
the chaparral shrub, *Ceanothus megacarpus*, after fire. - Ecology 61: 781-789,
1980.

7776 - SCHMUGGE, T.J.: Microwave approaches in hydrology. - Photogrammetric Eng.
Remote Sensing 46: 495-507, 1980.

7777 - SCHMUGGE, T.J., JACKSON, T.J., McKIM, H.L.: Survey of methods for soil moisture
determination. - Water Resour. Res. 16: 961-979, 1980.

7778 - SCHNABL, H.: Anion metabolism as correlated with the volume changes of guard
cell protoplasts. - Z. Naturforsch. 35: 621-626, 1980.

7779 - SCHNABL, H.: CO_2 and malate metabolism in starch-containing and starch-lacking
guard-cell protoplasts.- Planta 149: 52-58, 1980.

7780 - SCHNABL, H.: Der Anionenmetabolismus in stärkehaltigen und stärkefreien
Schliesszellprotoplasten. - Ber. Deut. bot. Ges. 93: 595-605, 1980.

7781 - SCHNABL, H., HAMPP, R.: *Vicia* guard cell protoplasts lack photosystem II activity. - Naturwissenschaften 67: 465-466, 1980.

7782 - SCHNABL, H., RASCHKE, K.: Potassium chloride as stomatal osmoticum in *Allium cepa* L., a species devoid of starch in guard cells. - Plant Physiol. 65: 88-93, 1980.

7783 - SCHNABL, H., SCHEURICH, P., ZIMMERMANN, U.: Mechanical stabilization of guard cell protoplasts of *Vicia faba*. - Planta 149: 280-282, 1980.

7784 - SCHNABL, H., VIENKEN, J., ZIMMERMANN, U.: Regular arrays of intramembranous particles in the plasmalemma of guard cell and mesophyll cell protoplasts of *Vicia faba*. - Planta 148: 231-237, 1980.

*7785 - SCHNETTER, R., HILGER., H.H., RICHTER, U.: Über den Bau des Perikarps der hygrochastischen Hülse von *Haematoxylum brasiletto* (*Caesalpiniaceae, Fabales*). - Bot. Jahrb. Syst. 101: 135-142, 1979.

7786 - SCHOBERT, B.: The importance of water activity and water structure during hyperosmotic stress in algae and higher plants. - Biochem. Physiol. Pflanz. 175: 175: 91-103. 1980.

7787 - SCHOCH, P.-G., ZINSOU, C., SIBI, M.: Dependence of the stomatal index on environmental factors during stomatal differentiation in leaves of *Vigna sinensis* L. 1. Effect of light intensity. - J. exp. Bot. 31: 1211-1216, 1980.

7788 - SCHONBECK, M.W., EGLEY, G.H.: Effects of temperature, water potential, and light on germination responses of redroot pigweed seeds to ethylene. - Plant Physiol. 65: 1149-1154, 1980.

*7789 - SCHONBECK, M.W., NORTON, T.A.: An investigation of drought avoidance in internal fucoid algae. - Bot. mar. 22: 133-144, 1979.

7790 - SCHONBECK, M.W., NORTON, T.A.: The effects on intertidal fucoid algae of exposure to air under various conditions. - Bot. mar. 23: 141-147, 1980.

7791 - SCHÖNHERR, J., ZIEGLER, H.: Water permeability of *Betula* periderm. - Planta 147: 345-354, 1980.

*7792 - SCHRÖDER, D.: Der Einfluss der Düngung auf den Wasserverbrauch von Rüben und Kartoffeln. - Kali-Briefe Fachgebiert 8: 1-9, 1976.

7793 - SCHULZE, E.-D., HALL, A.E. LANGE, O.L., EVENARI, M., KAPPEN, L., BUSCHBOM, U.: Long-term effects of drought on wild and cultivated plants in the Negev desert. I. Maximal rates of net photosynthesis. - Oecologia 45: 11-18, 1980.

7794 - SCHULZE, E.-D., LANGE, O.L., EVENARI, M., KAPPEN, L., BUSCHBOM, U.: Long-term effects of drought on wild and cultivated plants in the Negev desert. II. Diurnal patterns of net photosynthesis and daily carbon gain. - Oecologia 45: 19-25, 1980.

7795 - SCHUSTER, W., BRETSCHNEIDER-HERRMANN, B., MARQUARD, R.: Untersuchungen über den Einfluss von Temperatur, Tageslänge und Luftfeuchtigkeit auf die Qualität von Rapssamen. - Bodenkultur 31: 373-391, 1980.

7796 - SCHUURMAN, J.J.: Root and crop growth of oats as affected by the lenght of periods of high water-table. - Neth. J. agr. Sci. 28: 20-28, 1980.

7797 - SCIENZA, A., DÜRING, H.: Stickstoffernährung und Wasserhaushalt bei Reben. - Vitis 19: 301-307, 1980.

*7798 - SCOTT, N.S., MUNNS, R., BARLOW, E.W.R.: Polyribosome content in young and aged wheat leaves subjected to drought. - J. exp. Bot. 30: 905-911, 1979.

*7799 - SCOTTER, D.R., CLOTHIER, B.E., TURNER, M.A.: The soil water balance in Fragiaqualf and its effect on pasture growth in central New Zealand. - Aust. J. Soil Res. 17: 455-465, 1979.

*7800 - SEDIYAMA, G.C., PRUITT, W.O.: Estudo do microclima e dos perfis de umidade e dioxido de carbono no interior e acima do dossel vegetativo da cultura do sorgo (*Sorghum vulgare* Pers.) [Study of the microclimate and the humidity and carbon dioxide profiles in the interior and upper part of the vegetative canopy of the sorghum culture (*Sorghum vulgare* Pers.)] - Rev. Ceres 24: 563-570, 1977. [In Port, ab: E.]

*7801 - SEEMANN, J., CHIRKOV, Y.I., LOMAS, J., PRIMAULT, B.: Agrometeorology. - Springer-Verlag, Berlin - Heidelberg - New York 1979.

*7802 - SEEMANN, J.R., DOWNTON, W.J.S., BERRY, J.A.: Field studies of acclimation to high temperature: winter ephemerals in Death Valley. - Carnegie Inst. Year Book 78: 157-162, 1979.

.7803 - SEENAPPA, M., STOBBS, L.W., KEMPTON, A.G.: *Aspergillus* colonization of indian red pepper during storage. - Phytopathology 70: 218-222, 1980.

7804 - SEN, H., JANA, P.K., MAITY, S.P.: Effect of soil-moisture tensions and preceding winter crops on the yield and water-use efficiency of succeding crop of tossa jute. - Ind. J. agr. Sci. 50: 948-951, 1980.

*7805 - SEQUI, P., PETRUZZELLI, G.: La fertilizzazione dei terreni irrigui. [Fertilization of irrigated soils.] - Agrochimica 19: 34-58, 1975. [In Ital, ab: E,F, G,Span.]

*7806 - SETTER, T.L., BRUN, W.A., BRENNER, M.L.: Source/sink interactions in soybeans. I. A possible role of ABA, - Plant Physiol. 63 (Suppl.): 43, 1979.

7807 - SETTER, T.L., BRUN, W.A., BRENNER, M.L.: Stomatal closure and photosynthetic inhibition in soybean leaves induced by petiole girdling and pod removal. - Plant Physiol. 65: 884-887, 1980.

7808 - SETTER, T.L., BRUN, W.A., BRENNER, M.L.: Effect of obstructed translocation on leaf abscisic acid, and associated stomatal closure and photosynthesis decline. - Plant Physiol. 65: 1111-1115, 1980.

*7809 - SHANER, D.L., LYON, J.L.: Somatal cycling in *Phaseolus vulgaris* L. in response to glyphosate. - Plant Sci. Lett. 15: 83-87, 1979.

7810 - SHANER, D.L., LYON, J.L.: Interaction of glyphosate with aromatic amino acids on transpiration in *Phaseolus vulgaris*. - Weed Sci. 28: 31-35, 1980.

* 7811 - SHARKEY, T.D., RASCHKE, K.: Separation and measurement of direct and indirect effects of light on stomata. - Plant Physiol. 63 (Suppl.): 60, 1979.

7812 - SHARKEY, T.D., RASCHKE, K.: Effects of phaseic acid and dihydrophaseic acid on stomata and the photosynthetic apparatus. - Plant Physiol. 65: 291-297, 1980.

*7813 - SHARMA, G.K.: Cuticular features as indicators of environmental pollution. - USDA Forest Serv. gen. tech. Rep. NE-23 (Proceedings of the First Internationa Symposium on Acid Precipitation and the Forest Ecosystem) : 927-932, 1976.

7814 - SHARMA, G.K., CHANDLER, C., SALEMI, L.: Environmental pollution and leaf cuticular variation in Kudzu (*Pueraria lobata* Willd.) . - Ann. Bot. 45; 77-80, 1980.

*7815 - SHARMA, M.L., LUXMOORE, R.J.: Soil spatial variability and its consequences on simulated water balance. - Water Resour. Res. 15: 1567-1573, 1979.

7816 - SHARMA, R.N., BHARDWAJ, R.B.L.: Note on dry-matter accumulation and nitrogen uptake by durum and bread wheats under limited and adequate irrigation supply. - Ind. J. agr. Sci. 50: 877-879, 1980.

7817 - SHARMA, R.P., PARASHAR, K.S.: Effect of different levels of fertilizers and volumes of water applied at sowing time to winter season crops under dryland conditions. - Ind. J. Agron. 25: 169-175, 1980.

7819 - SHARP, R.R., YOCUM, C.F.: The kinetics of water exchange across the chloroplast membrane. - Biochem. biophys. Acta 592: 169-184, 1980.

7820 - SHAW, R.H., ROSS, K., MEYERS, C.: Evaluation of the management, yield, and water-use interactions on corn in northwestern Iowa. - Iowa State J. Res. 55: 119-126, 1980.

7821 - SHAYO-NGOWI, A., CAMPBELL, G.S.: Measurement of matric potential in plant tissue with a hydraulic press. - Agron. J. 72: 567-568, 1980.

7822 - SHCHERBAKOVA, A., KACPERSKA-PALACZ, A.: Modification of stress tolerance by dehydration pretreatment in winter rape hypocotyls. - Physiol. Plant. 48: 560-563, 1980.

*7823 - SHEORAN, I.S., BABBER, S., KHAN, M.I.: Water stress and starch accumulation in germinating guar (Cyanopsis tetragonaloba L.). - Plant Sci. Lett. 15: 159-163, 1979.

*7824 - SHEORAN, I.S., GARG, O.P.: Effect of chloride and sulphate salinity on germination and early seedling growth of moong. - Acta bot. Ind. 6 (Suppl.): 84-89, 1978.

7825 - SHEORAN, I.S., GARG, O.P.: Changes in isoenzymes of soluble malate dehydrogenase during germination of mung bean (Phaseolus aureus Roxb.) under salt stress. - Biol. Plant. 22: 384-387, 1980.

*7826 - SHEORAN, I.S., KHAN, M.I., GARG, O.P.: Effect of simulated drought and the amount of solution used on the germination and early seedling growth of guar (Cyanopsis tetragonoloba L.). - Forage Res. 5: 135-140, 1979.

7827 - SHERTZ, R.D., KENDER, J.W., MUSSELMAN, R.C.: Effects of ozone and sulfur dioxide on grapevines. - Scientia Hort. 13: 37-45, 1980.

7828 - SHIH, S.F., GASCHO, G.J.: Water requirement for sugarcane production. - Trans. ASAE 23: 934-937, 1980.

*7829 - SHIVE, J.B.,Jr., BROWN, K.W.: Quaking and gas exchange in leaves of cottonwood (Populus deltoides Michx.). - Plant Physiol. 59 (Suppl.) : 61: 1977.

*7830 - SHMAT'KO, I.G., SHAPOVAL, A.I., SHEVCHUK, N.V.: Ustoĭchivost' zelenykh pigmentov k vodnomu defitsitu i povyshennym temperaturam. [Resistance of green pigments to water deficit and elevated temperatures.] - In: Metody Otsenki Ustoĭchivosti Rasteniĭ k Neblagopriyatnym Usloviyam Sredy. Pp. 48-54. Kolos, Leningrad 1976. [In R.]

7831 - SHONE, M.G.T., BARTLETT, B.O., FLOOD, A.V.: A model of diffusion and mass flow of water in cylindrical membrane systems with application to plant roots. - J. Membrane Biol. 53: 171-177, 1980.

7832 - SHOUSE, P., JURY, W.A., STOLZY, L.H.: Use of deterministic and empirical models to predict potential evapotranspiration in an advective environment. - Agron. J. 72: 994-998, 1980.

7833 - SIDDIQUE, M.A., GOODWIN, P.B.: Seed vigour in bean (Phaseolus vulgaris L. cv. Apollo) as influenced by temperature and water regime during development and maturation. - J. exp. Bot. 31: 313-323, 1980.

7834 - SIDDIQUI, M.Q.: Some effects of rust infection and moisture stress on growth, diffusive resistance and distribution pattern of labelled assimilates in sunflower. - Aust. J. agr. Res. 31: 719-726, 1980.

7835 - SIGFRIDSSON, B.: Some effects of humidity on the light reaction of photosynthe-
       sis in the lichens *Cladonia impexa* and *Collema flaccidium*. - Physiol. Plant.
       49: 320-326, 1980.

7836 - SILK, W.K.: Plastochron indices in cantaloupe grown on an irrigation line
       source. - Bot. Gaz. 141: 73-78, 1980.

7837 - SILK, W.K., WAGNER, K.K.: Growth-sustaining water potential distributions in
       the primary corn root. A noncompartmented continuum model. -´Plant Physiol.
       66: 859-863, 1980.

7838 - SIMMONDS, J.: Increased seedling establishment of *Impatiens wallerana* in re-
       sponse to low temperature or polyethylene glycol seed treatments. - Can. J.
       Plant Sci. 60: 561-569, 1980.

7839 - ŠIMON, J.: Produkce ozimé a jarní pšenice na lehkých půdách pod vlivem závla-
       hy, výsevku a dusíkatého hnojení. [The production of winter and spring wheat
       on light-textured soils as influenced by irrigation, sowing rates, and nitro-
       genous fertilization.] - Rost. Výroba (Praha) 26: 621-630, 1980. [In Czech,
       ab: E,G,R.]

7840 - SINCLAIR, R.: Water potential and stomatal conductance of three *Eucalyptus*
       species in the Mount Lofty Ranges, South Australia: Responses to summer
       drought. - Aust. J. Bot. 28: 499-510, 1980.

7841 - SINCLAIR, T.R.: Leaf CER from post-flowering to senescence of field-grown
       soybean cultivars. - Crop Sci. 20: 196-200, 1980.

*7842 - SINGH, B.P.: Growth response of height isogenic barley to cultural treatments.
       - Fyton 37: 41-48, 1979.

7843 - SINGH, B.P., PHILLIPS, B.A.: Adaxial and abaxial stomatal frequency in deter-
       minate soybeans. - Fyton 38: 81-84, 1980.

*7844 - SINGH, G., SAXENA, G.S.: Comparative study on the effect of quality of irriga-
       tion water on the yield of barley, wheat and pea grown in soils of different
       texture. - Ind. J. agr. Res. 13: 199-202, 1979.

7845 - SINGH, N.T., PATEL, M.S., SINGH, R., VIG, A.C.: Effect of soil compaction on
       yield and water use efficiency of rice in a highly permeable soil. - Agron. J.
       72: 499-502, 1980.

7846 - SINGH, R., PRIHAR, S.S., SINGH, N.: Harvest-time moisture profiles and yield
       of unirrigated maize as influenced by variety and rainfall distribution. -
       Ind. J. agr. Sci. 50: 818-821, 1980

*7847 - SINGH, R.A., THAKUR, R.P.: Effect of presowing soaking treatments on germina-
       tion of soybean under moisture stress. - Ind. J. Ecol. 6: 98-102, 1979.

*7848 - SINGH, R.V.: Variation in moisture content in stems of plantation-grown *Chir*
       pine. - Aust. J. Forest Res. 8: 239-242, 1978.

7849 - SINGH, U.B., SINGH, R.M.: Effect of graded levels of moisture regimes, N and
       P fertilization on seed yield, oil content and NPK uptake by safflower. -
       Ind. J. Agron. 25: 9-17, 1980.

*7850 - SINHA, S.K., KHANNA, R.: Physiological, biochemical, and genetic basis of het-
       erosis. - Adv. Agron. 27: 123-174, 1975.

7851 - SIONIT, N., HELLMERS, H., STRAIN, B.R.: Growth and yield of wheat under CO_2
       enrichment and water stress. - Crop Sci. 20: 687-690, 1980.

7852 - SIONIT, N., TEARE, I.D., KRAMER, P.J.: Effects of repeated application of
       water stress on water status and growth of wheat. - Physiol. Plant. 50: 11-15,
       1980.

7853 - **SIPOŞ, G.**: Relatia dintre periodata de vegetaţie şi cerinţele culturilor agricole faţa de apǎ. [Relation between the vegetation period and the water requirements of the agricultural crops.] - An. Inst. Cercetari Cereale Plante
tehnice Fundulea 46: 261-266, 1980. [In Roum, ab: E,R.]

7854 - **SIPOŞ, G., CRISTEA, F.**: Cerinţele faţǎ de apǎ şi eficienţa udarilor la fasole.
[Bean water needs and irrigation efficiency.] - An. Inst. Cercetǎri Cereale Plant
Plante tehnice Fundulea 45: 337-341, 1980. [In Roum, ab: E,R.]

*7855 - **SIPOŞ, G., PǍLTINEANU, R.**: Consumul de apǎ şi regimul de irigare la principalele culturi agricole. [Water consumption and irrigation regime of principal
crop plants.] Producţia vegetala, Cereale, Plante tehnice 1975 (5): 3-9, 1975.
[In Roum.]

*7856 - **SISLER, E.C.**: Photobleaching of tobacco leaves. - Tobacco 178 (6): 55-59,
1976. Tobacco Sci. 20: 35-39, 1976.

*7857 - **SIVAKUMAR, M.V.K., VIRMANI, S.M.**: Measuring leaf-water potential in chickpea
with a pressure chamber. - Exp. Agr. 15: 377-383, 1979.

*7858 - **SIVAKUMARAN, S., HALL, M.A.**: Hormones in relation to stress recovery in *Populus robusta* cuttings. - J. exp. Bot. 30: 53-63, 1979.

7859 - **SIVAKUMARAN, S., HORGAN, R., HEALD, J., HALL, M.A.**: Effect of water stress on
metabolism of abscisic acid in *Populus robusta* X schnied and *Euphorbia lathy-
rus* L. - Plant Cell Environ 3: 163-173, 1980.

7860 - **SKAAR, H., JOHNSSON, A.**: Light induced transpiration in a chlorophyll deficient mutant of *Hordeum vulgare*. - Physiol. Plant. 49: 210-214, 1980.

7861 - **SKACHKOV, M.P., TRUKHAN, E.M.**: Fotoindutsirovannye izmeneniya diėlektricheskikh poter' v mikrovolnovoǐ oblasti u fotosinteziruyushchikh bakteriǐ *Rhodospirillum rubrum*. [Photoinduced changes of the dielectrical losses in microwave region in photosynthetic bacteria *Rhodospirillum rubrum*.] - Biofizika 25:
520-522, 1980. [In R, ab: E.]

7862 - **SKACHKOV, M.P., TRUKHAN, E.M., KHARCHENKO, S.G.**: Fotoindutsirovannye izmeneniya
diėlektricheskikh poter' v mikrovolnovoi oblasti v khromatoforakh fotosinteziruyushchikh bakteriǐ *Rhodospirillum rubrum*. [Photoinduced changes of dielectric
losses in the microwave region in chromatophores of photosynthesizing bacteria
Rhodospirillum rubrum.] - Biofizika 25: 564-566, 1980. [In R, ab: E.]

7863 - **SKJELVÅG, A.O.**: A crop-weather analysis model applied to field bean. - Int. J.
Biometeorol. 24: 301-313, 1980.

7864 - **SKOOG, F. (ed.)**: Plant Growth Substances 1979. Proceedings of the 10th International Conference on Plant Growth Substances, Madison, Wisconsin, July 22-
26, 1979. - Springer-Verlag, Berlin - Heidelberg - New York 1980.

7865 - **SLABBERS, P.J.**: Practical prediction of actual evapotranspiration. - Irrig.
Sci. 1: 185-196, 1980.

7866 - **SLACK, D.C., RIGGLE, F.R.**: Effects of Joule heating on thermocouple psychrometer water potential determinations. - Trans. ASAE 23: 877-883, 1980.

7867 - **SLAVÍK, L.**: Příspěvek ke studiu vláhových nároků ozimé pšenice jako podkladu
k volbě nejúčinnějšího závlahového režimu. [Study of moisture requirements of
winter wheat as a basis for choosing the best irrigation regime.] - Rost. Výroba (Praha) 26: 583-590, 1980. [In Czech, ab: E,G,R.]

7868 - **SLONOV, L.Kh., PETINOV, N.S.**: Soderzhanie nukleotidov i aktivnost' ATFaz v
list'yakh konopli v zavisimosti ot vodoobespechennosti. [Content of nucleoti-
des and activity of ATPases in hemp leaves in relation to water supply.] -
Fiziol. Rast. 27: 1095-1100, 1980. [In R, ab: E.]

#7869 - SMITH, D.C.: Symbiosis and the biology of lichenised fungi. - In: JENNINGS, D.
        H., LEE, D.L. (ed.): Symbiosis. Symposia of the Society for Experimental Bio-
        logy. No. 29. Pp. 373-405. Cambridge University Press, New York - Cambridge
        1975.

7870 - SMITH, G.S., MIDDLETON, K.R., EDMONDS, A.S.: Sodium nutrition of pasture
        plants. I. Translocation of sodium and potassium in relation to transpiration
        rates. - New Phytol. 84: 603-612, 1980.

7871 - SMITH, J.A.C., MILBURN, J.A.: Osmoregulation and the control of phloem-sap
        composition in *Ricinus communis* L. - Planta 148: 28-34, 1980.

7872 - SMITH, J.A.C., MILBURN, J.A.: Phloem transport, solute flux and the kinetics
        of sap exudation in *Ricinus communis* L. - Planta 148: 35-41, 1980.

7873 - SMITH, J.A.C., MILBURN, J.A.: Phloem turgor and the regulation of sucrose
        loading in *Ricinus communis* L. - Planta 148: 42-48, 1980.

7874 - SMITH, J.A.C., MILBURN, J.A.: Water stress and phloem loading. - Ber. Deut.
        bot. Ges. 93: 269-280, 1980.

*7875 - SMITH, M.W., KENWORTHY, A.L., BEDFORD, C.L.: The response of fruit trees to
        injection of nitrogen through a trickle irrigation system. - J. Amer. Soc.
        hort. Sci. 104; 311-313, 1979.

7876 - SMITH, W.K.: Importance of aerodynamic resistance to water use efficiency in
        three conifers under field conditions. - Plant Physiol. 65: 132-135, 1980.

7877 - SMITH, W.K., GELLER, G.N.: Leaf and environmental parameters influencing
        transpiration: Theory and field measurements. - Oecologia 46: 308-313, 1980.

*7878 - SMITTLE, D.A., WILLIAMSON, R.E.: Effect of soil compaction on nitrogen and
        water use efficiency, root growth, yield, and fruit shape of pickling cucum-
        bers. - J. Amer. Soc. hort. Sci. 102: 822-825, 1977.

7879 - SOBRADO, M.A., MEDINA, E.: General morphology, anatomical structure, and nu-
        trient content of sclerophyllous leaves of the 'bana' vegetation of Amazonas.
        - Oecologia 45: 341-345, 1980.

7880 - SOER, G.J.R.: Estimation of regional evapotranspiration and soil moisture con-
        ditions using remotely sensed crop surface temperatures. - Remote Sensing
        Environ. 9: 27-45, 1980.

7881 - SOJKA, R.E., STOLZY, L.H.: Soil-oxygen effects on stomatal response. - Soil
        Sci. 130: 350-358, 1980.

7882 - SOLÁROVÁ, J.: Diffusive conductances of adaxial (upper) and abaxial (lower)
        epidermes: Response to quantum irradiance during development of primary *Pha-
        seolus vulgaris* L. leaves. - Photosynthetica 14: 523-531, 1980.

7883 - SOLÁROVÁ, J., POSPÍŠILOVÁ, J., TICHÁ, I., ČATSKÝ, J., PLESKANKA, J.: Modifika-
        ce závislosti epidermální vodivosti na kvantové ozářenosti biologickými a eko-
        logickými faktory. [Relationship between epidermal conductance and quantum
        irradiance modified by biological and ecological factors.] - In: Dny Rostlinné
        Fyziologie II. Pp. 375-379. Vysoká Škola Zemědělská, Brno 1980. [In Czech, ab:
        E,R.]

7884 - SORENSEN, V.M., HANKS, R.J., CARTEE, R.L.: Cultivation during early season
        and irrigation influences on corn production. - Agron. J. 72: 266-270, 1980.

*7885 - SPALDING, M.H., STUMPF, D.K., KU, S.B., EDWARDS, G.E.: Diurnal variations in
        internal CO_2 and O_2 concentrations in *Sedum praealtum* in relation to Crassula-
        cean acid metabolism. - Plant Physiol. 63 (Suppl.): 63, 1979.

*7886 - ŠPÁNIK, F., KRAJČÍROVÁ, Z.: Vzťah využiteľnej pôdnej vody k rastu organickej
hmoty ozimej pšenice Uhřiněvská-2. [Relation of soil water availability to
organic matter production of winter wheat cv. Uhřiněvská-2.] - In: Bilancia
Energie a Vody v Polných a Lesných Ekosystémoch. Pp. 59-69. Vysoká Škola Poľnohospodárská, Nitra 1979. [In Slov.]

7887 - SPEARS, B.M., ROSE, S.T., BELLES, W.S.: Effect of canopy cover, seeding depth,
and soil moisture on emergence of *Centaurea maculosa* and *C. diffusa*. - Weed
Res. 20: 87-90, 1980.

*7888 - SPEKTOROV, K.S., STROGONOV, B.P.: Mekhanizmy, obespechivayushchie ustoĭchivost'
morskikh i presnovodnykh vodoroslei k izmeneniyu osmoticheskogo davleniya
okruzhayushchei sredy. [Mechanisms of marine and freshwater algae tolerance to
changes in osmotic pressure of the external medium.] - Fiziol. Rast. 26: 967-
977, 1979. [In R, ab: E.]

7889 - SPEKTOROV, K.S., STROGONOV, B.P.: Adaptatsiya morskoĭ odnokletochnoĭ vodorosli
Dunaliella primolecta k ponizhennomu osmoticheskomu davleniyu i nizkomu soder-
zhaniyu NaCl v srede. [Adaptation of unicellular marine alga *Dunaliella primolecta* to low osmotic pressure and NaCl content in the medium.] - Fiziol. Rast.
27: 259-265, 1980. [In R, ab: E.]

7890 - SPIERS, J.M.: Influence of peat moss and irrigation on establishment of "Tif-
blue" blueberry. - Mississippi agr. Forest exp. Sta. Res. Rep. 4 (18): 1-4,
1980.

7891 - SPYROPOULOS, C.G., LAMBIRIS, M.P.: Effect of water stress on germination and
reserve carbohydrate metabolism in germinating seeds of *Ceratonia siliqua* L.
- J. exp. Bot. 31: 851-857, 1980.

*7892 - STAMBOLIEV, M., IGNATOVA, A.: Vliyanie na narastvashchite normi na azotno,
fosforno i kalievo torene v"rkhu chimichniya s"stav i dobiva ot pshenitsata
za raĭona na severozapadna B"lgariya. [Effect of increasing rates of nitrogen,
phosphorus and potassium fertilizer application on the chemical composition of
and yields from wheat in the area of north-western Bulgaria.] - Pochvoznanie
Agrochim. 11 (2): 61-72, 1976. [In Bulg, ab: E,R.]

*7893 - STĂNCIULESCU, G., PODOLEANU, M., IANOŞI, S.: Efectul combinat al regimului de
irigare şi fertilizare asupra unor însuşiri de producţie morfo-fiziologice la
hibrizii de tomate timpurii. [The combined effect of irrigation and fertilization conditions on some yield and morpho-physiological features of the early
tomato hybrids.] - An. Inst. Cercetări Pentru Leguminocult. Floricult. 5: 227-
239, 1979. [In Roum, ab: E,F.]

*7894 - STANHILL, G.: Cotton. - In: MONTEITH, J.L. (ed.): Vegetation and the Atmos-
phere. Vol. 2. Case Studies. Pp. 121-150. Academic Press - London - New York -
San Francisco 1976.

7895 - STAŇKOVÁ, J.: Vodní bilance listů rododendronů. [Water balance of the rhodo-
dendron leaves.] - In: Dny Rostlinné Fyziologie II. Pp. 380-385. Vysoká Škola
Zemědělská, Brno 1980. [In Czech, ab: E,R.]

7896 - STANLEY, C.D., KASPAR, T.C., TAYLOR, H.M.: Soybean top and root response to
temporary water tables imposed at three different stages of growth. - Agron.
J. 72: 341-346, 1980.

7897 - STANSELL, J.R., SMITTLE, D.A.: Effects of irrigation regimes on yield and
water use of snap bean (*Phaseolus vulgaris* L.). - J. Amer. Soc. hort. Sci.
105: 869-873, 1980.

7898 - STARK, J.C., JARRELL, W.M.: Salinity-induced modifications in the response of
maize to water deficits. - Agron. J. 72: 745-748, 1980.

7899 - STATLER, G.D., NORDGAARD, J.T.: Leaf wettability of wheat in relation to infection by *Puccinia recondita* f. sp. *tritici*. - Phytopathology 70: 641-643, 1980.

*7900 - STEELE SCOTT, N., MUNNS, R., BARLOW, E.W.R.: Polyribosome content in young and aged wheat leaves subjected to drought. - J. exp. Bot. 30: 905-911, 1979.

*7901 - STEHLÍK, K.: Využitelnost srážek v bilanci vody zavlažovaných kultur. [Water consumption in water balance of irrigated crops.] - In: Bilancia Energie a Vody v Polných a Lesných Ekosystémoch. Pp. 79-87. Vysoká Škola Poľnohospodárská, Nitra 1979. [In Czech.]

7902 - STEHLÍK, K., MUSIL, J.: Závlaha jetelotrávy škrobárenskými odpadními vodami z výnosového hlediska. [Irrigation of clover-grass mixture with starch factory waste waters as affecting the yields.] - Rost. Výroba (Praha) 26: 591-598, 1980. [In Czech, ab: E,G,R.]

7903 - STELZER, R., LAÜCHLI, A.: Salt- and flooding tolerance of *Puccinellia peisonis*. IV. Root respiration and the role of aerenchyma in providing atmospheric oxygen to the roots. - Z. Pflanzenphysiol. 97: 171-178, 1980.

7904 - STEPONKUS, P.L.: A unified concept of stress in plants? - In: RAINS, D.W., VALENTINE, R.C., HOLLAENDER, A. (ed.): Genetic Engineering of Osmoregulation. Impact on Plant Productivity for Food, Chemicals and Energy. Pp. 235-255. Plenum Press, New York - London 1980.

7905 - STEPONKUS, P.L., CUTLER, J.M., O'TOOLE, J.C.: Adaptation to water stress in rice. - In: TURNER, N.C., KRAMER, P.J. (ed.): Adaptation of Plants to Water and High Temperature Stress. Pp. 401-418. John Wiley & Sons, New York - Chichester - Brisbane - Toronto 1980.

7906 - STEUDLE, E., SMITH, J.A.C., LÜTTGE, U.: Water-relation parameters of individual mesophyll cells of the crassulacean acid metabolism plant *Kalanchoë daigremontiana*. - Plant Physiol. 66: 1155-1163, 1980.

7907 - STEWART, C.R., HANSON, A.D.: Proline accumulation as a metabolic response to water stress. - In: TURNER, N.C., KRAMER, P.J. (ed.): Adaptation of Plants to Water and High Temperature Stress. Pp. 173-189. John Wiley & Sons, New York - Chichester - Brisbane - Toronto 1980.

*7908 - STIGTER, C.J., WELGRAVEN, A.D.: An improved radiation protected differential thermocouple psychrometer for crop environment. - Arch. Meteorol. Geophys. Bioclimatol., Ser. B 24: 177-187, 1976.

*7909 - STOCK, H.-G.: Vergleichende Bodenfeuchtemessungen mit Tensiometern und Bohrstock auf einem Lö4-Standort. - Arch. Acker- Pflanzenbau Bodenk. 22: 611-614, 1978.

*7910 - STOCK, H.-G., WICKE, H.-J., MÜLLER, C.: Bodenfeuchteansprüche von Getreide in verschiedenen Entwicklungsstadien. - Arch. Acker- Pflanzenbau Bodenk. 20: 791 -805, 1976.

*7911 - STOFFERS, A.L.: Ecological aspects of the vegetation of Curaçao, Netherlands Antilles. 1. Water relations. - In: LARSEN, K., HOLM-NIELSEN, L.B. (ed.): Tropical Botany. Pp. 251-262. Academic Press, London 1979.

*7912 - STOKER, R.: Increasing pea yields with irrigation. - N. Zeal. J. Agr. 135 (3): 23-26, 1977.

*7913 - STOKER, R.: Yield and water use of sweet lupins. - Proc. Agron. Soc. N. Zeal. 8: 23-26, 1978.

*7914 - STOKER, R., DREWITT, E.G.: Cropping on light land with irrigation. - N. Zeal. J. Agr. 135 (6): 34-37, 1977.

7915 - STONE, D.A.: Effects of textural amendment of coarse soils on crop growth and water use. - J. Sci. Food Agr. 31: 769-776, 1980.

7916 - STONE, D.A., ROWSE, H.R.: Effects of textural amendment of coarse soils on seed-bed water content and seedling emergence. - J. Sci. Food Agr. 31: 759-768, 1980.

*7917 - STONE, J.E., MARX, D.B., DOBRENZ, A.K.: Interaction of sodium chloride and temperature on germination of two alfalfa cultivars. - Agron. J. 71: 425-427, 1979.

*7918 - STONE, J.F. (ed.): Plant Modification for More Efficient Water Use. - Elsevier Scientific Publishing Company, Amsterdam - Oxford - New York 1975.

*7919 - STONE, L.F., DA SILVEIRA, P.M., De OLIVEIRA, A.B., DE AQUINO, A.R.L.: Efeitos da supressão de água em diferentes fases do crescimento na produção do arroz irrigado. [Effect of whithholding water at different growth stages on the flooded rice yield.] - Pesqui agropec. Bras. 14: 105-109, 1979. [In Port, ab: E.]

*7920 - STÖPEL, W., BECHSTÄDT, O., SCHWARZ, K.: Untersuchungen zur Wirkung steigender Abwassergaben auf Ertrag und Qualität von Welschem Weidengras, Zückerrübe und Winterweizen auf schweren Böden. - Arch. Acker- Pflanzenbau Bodenk. 19: 577-586, 1975.

7921 - STORDEUR, R.: Einfluss der im Strassenwinterdienst eingesetzten $MgCl_2$-Sole auf das ökologische Verhalten von *Puccinellia distans* (Jacq.) Parl. und *Lolium perenne* L. Wirkung auf Keimung, Wachstum, Konkurrenzverhalten und potentielle osmotische Drücke. - Flora 170: 271-289, 1980.

7922 - STOUT, D.G.: Alfalfa water status and cold hardiness as influenced by cold acclimation and water stress. - Plant Cell Environ. 3: 237-241, 1980.

7923 - STOUT, D.G., SIMPSON, G.M., FLOTRE, D.M.: Drought resistance of *Sorghum bicolor* L. Moench. 3. Seed germination under osmotic stress. - Can. J. Plant Sci. 60: 13-24, 1980.

*7924 - STOYANOV, Zh., DAKEV, T.: Vliyanie na mineralnoto khranene v"rkhu transpiratsiyata na d"rvesni fidanki pri estestvena obstanovka. [Effect of mineral nutrition on tree seedling transpiration under natural conditions.] - Fiziol. Rast. (Sofia) 2 (3): 77-82, 1976. [In Bulg, ab: E,R.]

7925 - STRONG, W.M., BARRY, G.: The availability of soil and fertilizer phosphorus to wheat and rape at different water regimes. - Aust. J. Soil Res. 18: 353-362, 1980.

7926 - STUMPE, H.: Einfluss des Vorfeuchtertrages und der Beregnung der Vorfrucht auf die im Boden verbleitenden mengen an anorganischem Stickstoff. - Arch. Acker- Pflanzenbau Bodenk. 24: 233-240, 1980.

7927 - STUTTE, C.A., WEILAND, R.T.: Irrigation and temperature influences on soybean yields. - Arkansas Farm Res. 29 (3): 8, 1980.

*7928 - SUAREZ, J.J., HERNANDEZ, A.: Effect of soil water deficit on the nutrient uptake by *Glycine wightii*. - Cuban J. agr. Sci. 13: 315-322, 1979.

*7929 - SUBRAMANYAM, P., VIJAYA KUMAR, C.S.K., RAO, A.S.: Transpiration in rust infected leaves of groundnut. - Curr. Sci. 47: 302-304, 1978.

*7930 - SUDNITSYN, I.I., EGOROV, Yu.V., SIDOROVA, M.A.: Optimizatsiya vodnogo rezhima pochv. [Optimization of soil water regime.] - Nauch. Dokl. vyssh. Shkoly, biol. Nauki 20 (11): 127-136, 1977. [In R.]

*7931 - SUH, H.W., DAYTON, A.D., CASADY, A.J., LIANG, G.H.: Diallel cross analysis of stomatal density and leaf-blade area in grain sorghum, *Sorghum bicolor.* - Can. J. Genet. Cytol. 18: 679-686, 1976.

7932 - SUMAYAO, C.R., KANEMASU, E.T., BRAKKE, T.W.: Using leaf temperature to assess evapotranspiration and advection. - Agr. Meteorol. 22: 153-166, 1980.

7933 - SUNDQVIST, C., BJÖRN, L.O., VIRGIN, H.I.: Factors in chloroplast differentiation. - In: REINERT, J. (ed.): Chloroplasts. Pp. 201-224. Springer-Verlag, Berlin - Heidelberg - New York 1980.

*7934 - SURÁNYI, D.: Transpiration of apricot trees in connection with apoplectic dieback. - Acta phytopathol. Acad. Sci. Hung. 14: 399-407, 1979.

7935 - SVESHNIKOVA, V.M., CHOIJZHAMTZ, B.: Vodnyǐ rezhim nekotorykh vidov roda *Stipa* v Mongolii. [Water balance in some species of *Stipa* in Mongolia.] - Bot. Zh. 65: 816-825, 1980. [In R.]

*7936 - SWANSON, R.H., BENECKE, U., HAVRANEK, W.M.: Transpiration in mountain beech estimated simultaneously by heat-pulse velocity and climatised cuvette. - N. Zeal. J. Forest Sci. 9: 170-176, 1979.

7937 - SYBER, A.Yu.: Reaktsii ust'its na izmeneniya vlazhnosti vozdukha i uchastie CO_2 v ètikh reaktsiyakh. [Stomatal responses to air humidity changes and the role of carbon dioxide in these responses.] - Fiziol. Rast. 27: 515-524, 1980. [In R, ab: E.]

*7938 · SYKORA, K.V.: The effects of the severe drought of 1976 on the vegetation of some moorland pools in the Netherlands. - Biol. Conserv. 16: 145-162, 1979.

*7939 - SZAREK, S.R., WOODHOUSE, R.M.: Ecophysiological studies of Sonoran Desert plants. I. Diurnal photosynthesis patterns of *Ambrosia deltoidea* and *Olneya tesota.* - Oecologia 26: 225-234, 1976.

*7940 - TAIZ, L., MÉTRAUX, J.-P.: The kinetics of bidirectional growth of stem sections from etiolated pea seedlings in response to acid, auxin and fusicoccin. - Planta 146: 171-178, 1979.

7941 - TAKEBA, G.: Accumulation of free amino acids in the tips of non-thermodormant embryonic axes accounts for the increase in the growth potential of New York lettuce seeds. - Plant Cell Physiol. 21: 1639-1644, 1980.

7942 - TAKEDA, T., HAKOYAMA, S., AGATA, W., FURUYA, S.: [Studies on weed vegetation in non-cultivated paddy fields. III. Effects of different soil moisture levels on growth during early stage of some summer grasses.] - Jap. J. Crop Sci. 49: 432-438, 1980. [In Jap, ab: E.]

7943 - TAN, C.S., FULTON, J.M.: Ratio between evapotranspiration of irrigated crops from floating lysimeters and class A pan evaporation. - Can. J. Plant Sci. 60: 197-201, 1980.

*7944 - TANZARELLA, O.A., BLANCO, A.: Stomatal frequency and size in durum wheat. - Genet. agr. 33: 355-362, 1979.

*7945 - TARGON, P.G., DOBROVOL'SKAYA, M.G., GRUBAYA, Zh.F., MAKOVETSKAYA, R.V.: Vodouderzhivayushchaya sposobnost' nekotorykh drevesnykh porod, introdutsirovannykh v Moldavii, i ee zavisimost' ot obmena veshchestv. [Water holding capacity of some tree species introduced in Moldavia and dependence of it on metabolism.] - Nauch. Dokl. vyssh. Shkoly, biol. Nauki 20 (12): 103-107, 1977. [In R.]

7946 - TAYLOR, F.E., COBB, A.H., DAVIES, L.G.: The effect of bentazone on stomatal behaviour in *Chenopodium album* L. - New Phytol. 85: 369-376, 1980.

7947 - TAYLOR, H.M.: Modifying root systems of cotton and soybean to increase water absorption. - In: TURNER, N.C., KRAMER, P. (ed.): Adaptation of Plants to Water and High Temperature Stress. Pp. 75-84. John Wiley & Sons, New York - Chichester - Brisbane - Toronto 1980.

7948 - TAYLOR, H.M.: Postponement of severe water stress in soybeans by rooting modifications: A progress report. - In: CORBIN, F.T. (ed.): World Soybean Research Conference II: Proceedings. Pp. 161-178. Westview Press, Boulder 1980.

7949 - TAYLOR, H.M.: Soybean growth and yield as affected by row spacing and by seasonal water supply. - Agron. J. 72: 543-547, 1980.

7950 - TAYLOR, J.A., WEST, D.W.: The use of Evan's blue stain to test the survival of plant cells after exposure to high salt and high osmotic pressure. - J. exp. Bot. 31: 571-576, 1980.

7951 - TAZAKI, T., ISHIHARA, K., USHIJIMA, T.: Influence of water stress on the photosynthesis and productivity of plants in humid areas. - In: TURNER, N.C., KRAMER, P.J. (ed.): Adaptation of Plants to Water and High Temperature Stress. Pp. 309-321. John Wiley & Sons, New York - Chichester - Brisbane - Toronto 1980.

7952 - TAZAWA, M., SHIMMEN, T.: Direct demonstration of the involvement of chloroplasts in the rapid light-induced potential change in tonoplast-free cells of *Chara australis*. Replacement of *Chara* chloroplasts with spinach chloroplasts. - Plant Cell Physiol. 21: 1527-1534, 1980.

7953 - TEGLER, B., KERSHAW, K.A.: Studies on lichen-dominated systems. XXIII. The control of seasonal rates of net photosynthesis by moisture, light, and temperature in *Cladonia rangiferina*. - Can. J. Bot. 58: 1851-1858, 1980.

7954 - TENHUNEN, J.D., LANGE, O.L., BRAUN, M., MEYER, A., LÖSCH, R., PEREIRA, J.S.: Midday stomatal closure in *Arbutus unedo* leaves in a natural macchia and under simulated habitat conditions in an environmental chamber. - Oecologia 47: 365-367, 1980.

7955 - TENHUNEN, J.D., LANGE, O.L., HARLEY, P.C., MEYER, A., GATES, D.M.: The diurnal time course of net photosynthesis of soybean leaves: Analysis with a physiologically based steady-state photosynthesis model. - Oecologia 46: 314-321, 1980.

*7956 - TENHUNEN, J.D., MEYER, A., LANGE, O.L., GATES, D.M.: Field test of a physiologically based steady-state photosynthesis model for whole leaves (WHOLEPHOT). - Plant Physiol 63 (Suppl.): 122, 1979.

7957 - TENHUNEN, J.D., MEYER, A., LANGE, O.L., GATES, D.M.: Development of a photosynthesis model with an emphasis on ecological applications. V. Test of the applicability of a steady-state model to description of net photosynthesis of *Prunus armeniaca* under field conditions. - Oecologia 45: 147-155, 1980.

*7958 - TEODORESCU, G.: Cercetări privind metodele de irigare a legumelor cultivate în solariile tip tunel. [Research on the irrigation methods of the vegetables grown under plastic tunnels.] - An. Inst. Cercetări Pentru Leguminocult. Floricult. 5: 89-100, 1979. [In Roum, ab: E,F.]

*7959 - TEODORESCU, G.: Regimul de irigare le tomate, ardei şi vinete cultivate în solar. [Irrigation conditions for the tomatoes, peppers and egg-plants grown in solariums.] - An. Inst. Cercetări Pentru Leguminocult. Floricult. 5: 101-112, 1979. [In Roum, ab: E,F.]

7960 - TERAMURA, A.H.: Relationships between stand age and water repellency of chaparral soils. - Bull. Torrey bot. Club 107: 42-46, 1980.

7961 - TERAMURA, A.H., BIGGS, R.H., KOSSUTH, S.: Effects of ultraviolet-B irradiances on soybean. II. Interaction between ultraviolet-B and photosynthetically active radiation on net photosynthesis, dark respiration, and transpiration. - Plant Physiol. 65: 483-488, 1980.

7962 - TERAO, H., INOUYE, J.: Effect of low water potential of the culture medium on mesocotyl elongation of rice seedlings. - Plant Cell Physiol. 21: 1661-1666, 1980.

*7963 - TEŞU, C., MERLESCU, E., PÎNZARU, D., AVARVAREI, I., TÎNJALĂ, I.: Changes, by irrigation, of physical, chemical and morphological properties in leached chernozem of Podu-Iloaiei. - Lucrări Stiint. Inst. agr. "Ion Ionescu de la Brad" Iaşi, Ser. Agr. 1977: 3-6, 1977.

*7964 - TEŞU, V., TOMA, L.-D., COSMULESCU, E., MERLESCU, E.: Influenţa diferitelor grade de salinitate asupra capacităţii de absorbţie a energiei luminoase si a elementelor minerale la soiul de grîu Aurora şi Dacia. [The influence of different salinity degrees on absorption capacity of bright energy and of mineral elements in Aurora and Dacia wheat varieties.] - Lucrări Stiint. Inst. agr. "Ion Ionescu de la Brad" Iaşi, Ser. Agr. 1977: 23-24, 1977. [In Roum.]

7965 - THAKUR, P.S., RAI, V.K.: Water stress effects on maize: carbohydrate metabolism of resistant and susceptible cultivars of Zea mays L. - Biol. Plant. 22: 50-56, 1980.

7966 - THALOUARN, P., HELLER, R.: Influence des sels mercuriques sur les mouvements d'eau et de solutés de disques de Betterave rouge. - Physiol. vég. 18: 289-299, 1980.

7967 - THANKI, Y.J., KOTHARI, I.L.: Structure and development of endocarpic stomata in the fruit of Momordica dioica. - Flora 169: 351-354, 1980.

*7968 - THIMANN, K.V., SATLER, S.O.: Relation between leaf senescence and stomatal aperture. - Plant Physiol. 63 (Suppl.): 72: 1979.

*7969 - THIMMEGOWDA, S.: Relative effects of soil moisture regimes and fertilizer levels on growth and yield of Lerma Rojo wheat. - Madras agr. J. 62: 559-563, 1975.

*7970 - THIMMEGOWDA, S., PUTTASWAMY, S., SHANTHAMALLAIAH, N.R., KRISHNAMURTHY, K.: Effect of soil moisture stress on growth and yield of finger millet. - Ind. J. Agron. 21: 358-360, 1976.

7971 - THOMAS, J.R.: Osmotic and specific salt effects on growth of cotton. - Agron. J. 72: 407-412, 1980.

*7972 - THOMAS, S., BIRD, I.F., CORNELIUS, M.J., KEYS, A.J.: Photosynthesis, photorespiration and productivity: Measurement of photorespiration in field crops. - In: COOMBS, J. (ed.): 4th International Congress on Photosynthesis. Pp. 380-381. UKISES, London 1977.

7973 - THOMAS, T.E., TURPIN, D.H.: Desiccation enhanced nutrient uptake rates in the intertidal alga Fucus distichus. - Bot. Mar. 23: 479-481, 1980.

7974 - THORPE, N.: Accumulation of carbon compounds in the epidermis of five species with either different photosynthetic systems or stomatal structure. - Plant Cell Environ. 3: 451-460, 1980.

7975 - THULLEN, R.J., KEELEY, P.E.: Competition between yellow nutsedge (Cyperus esculentus) and Japanese millet (Echinochloa crus-galli var. frumentacea). - Weed Sci. 28: 24-26, 1980.

*7976 - TING, I.P., LaPRE, L., LAZZARO, C.: Effect of reduced oxygen on gross CO_2 uptake. - Plant Physiol. 63 (Suppl.): 39, 1979.

☆7977 - **TKACHUK, K.S., ROZHKO, I.I.:** Uptake and metabolism of ^{15}N and ^{14}C in winter wheat under different soil moisture. - In: KUDREV, T., STOYANOV, I., GEORGIE-VA, V. (ed.): Mineral Nutrition of Plants. Vol. II. Pp. 256-262. Publishing House Central Cooperative Union, Sofia 1979.

7978 - **TOMAR, V.S., O'TOOLE, J.C.:** Design and testing of a microlysimeter for wetland rice. - Agron. J. 72: 689-692, 1980.

7979 - **TOMAR, V.S., O'TOOLE, J.C.:** Water use in lowland rice cultivation in Asia: A review of evapotranspiration. - Agr. Water Manage. 3: 83-106, 1980.

7980 - **TOPP, G.C., DAVIS, J.L., ANNAN, A.P.:** Electromagnetic determination of soil water content: Measurements in coaxial transmission lines. - Water Resour. Res. 16: 574-582, 1980.

7981 - **TOPP, G.C., ZEBCHUK, W.D., DUMANSKI, J.:** The variation of *in situ* measured soil water properties within soil map units. - Can. J. Soil Sci. 60: 497-509, 1980.

☆7982 - **TREICHEL, S.:** Der Einfluss von NaCl auf den Prolinstoffwechsel bei Halophyten. - Ber. Deut. bot. Ges. 92: 73-85, 1979.

7983 - **TROUGHT, M.C.T., DREW, M.C.:** The development of waterlogging damage in wheat seedlings (*Triticum aestivum* L.). I. Shoot and root growth in relation to changes in the concentrations of dissolved gases and solutes in the soil solution. - Plant Soil 54: 77-94, 1980.

7984 - **TROUGHT, M.C.T., DREW, M.C.:** The development of waterlogging damage in wheat seedlings (*Triticum aestivum* L.). II. Accumulation and redistribution of nutrients by the shoot. - Plant Soil 56: 187-199, 1980.

7985 - **TROUGHTON, A.:** Production of root axes and leaf elongation in perennial ryegrass in relation to dryness of the upper soil layer. - J. agr. Sci. 95: 533-538, 1980.

7986 - **TRUKSA, J., ZÁBORSKÝ, J.:** Úrodnost' hybridov kukurice pri rôznej organizácii porastu v závlahových podmienkach. [The yield of maize hybrids under different organization schemes of irrigated stands.] - Rost. Výroba (Praha) 26: 639-649, 1980. [In Slov, ab: E,G,R.]

☆7987 - **TSCHAKALOVA, E.:** Struktur-Funktionsbeziehungen bei *Phaseolus vulgaris* L. unter besonderer Berücksichtigung der Photosynthese. I. Untersuchung der Spaltöffnungen und des Interzellularvolumens von *Phaseolus vulgaris*- Blättern im Verlauf der Ontogenese. - Godishnik Sofii. Univ., biol. Fak. Kn. 2, 68: 1-10, 1976.

☆7988 - **TSUNODA, S.:** Characteristic of photosynthesis and environmental adaptation of rice. - In: WU, H.P., HSIEH, K.C. (ed.): Proceedings of the ROC - Japan Symposium on Rice Productivity. Inst. Bot. Acad. Sinica Monograph, Ser. 3. Pp. 3-8. Taiwan 1979.

7989 - **TUCKER, C.J.:** Remote sensing of leaf water content in the near infrared. - Remote Sensing Environ. 10: 23-32, 1980.

7990 - **TURK, K.J., HALL, A.E.:** Drought adaptation of cowpea. II. Influence of drought on plant water status and relations with seed yield. - Agron. J. 72: 421-427, 1980.

7991 - **TURK, K.J., HALL, A.E.:** Drought adaptation of cowpea. III. Influence of drought on plant growth and relations with seed yield. - Agron. J. 72: 428-433, 1980.

7992 - **TURK, K.J., HALL, A.E.:** Drought adaptation of cowpea. IV. Influence of drought on water use, and relations with growth and seed yield. - Agron. J. 72: 434-439, 1980.

7993 - TURK, K.J., HALL, A.E., ASBELL, C.W.: Drought adaptation of cowpea. I. Influence of drought on seed yield. - Agron. J. 72: 413-420, 1980.

7994 - TURNER, D.W., BARKUS, B.: Plant growth and dry-matter production of the 'Williams' banana in relation to supply of potassium, magnesium and manganese in sand culture. - Scientia Hort. 12: 27-45, 1980.

7995 - TURNER, N.C., JONES, M.M.: Turgor maintenance by osmotic adjustment: A review and evaluation. - In: TURNER, N.C., KRAMER, P.J. (ed.): Adaptation of Plants to Water and High Temperature Stress. Pp. 87-103. John Wiley & Sons, New York - Chichester - Brisbane - Toronto 1980.

7996 - TURNER, N.C., KRAMER, P.J. (ed.): Adaptation of Plants to Water and High Temperature Stress. - John Wiley & Sons, New York - Chichester - Brisbane - Toronto 1980.

7997 - TURNER, N.C., LONG, M.J.: Errors arising from rapid water loss in the measurement of leaf water potential by the pressure chamber technique. - Aust. J. Plant Physiol. 7: 527-537, 1980.

*7998 - TUŽINSKÝ, L.: Bilancia pôdnej vody vo vybraných lesných ekosystémoch Slovenska. [Soil water balance in representative forest ecosystems in Slovakia.] - In: Bilancia Energie a Vody v Polných a Lesných Ekosystémoch. Pp. 125-134. Vysoká Škola Polnohospodárská, Nitra 1979. [In Slov.]

*7999 - TVERGYAK, P.J., RICHARDSON, D.G.: Diurnal changes of leaf and fruit water potentials of sweet cherries during the harvest period. - HortScience 14: 520-521, 1979.

8000 - TYREE, M.T., YIANOULIS, P.: The site of water evaporation from sub-stomatal cavities, liquid path resistances and hydroactive stomatal closure. - Ann. Bot. 46: 175-193, 1980.

*8001 - TYSHKEVICH, G.L.: Biologo-fiziologicheskie osobennosti raznykh vidov i ėkotipov buka. [Biological and physiological characteristics of different beach species and ecotypes.] - Nauch. Dokl. vyssh. Shkoly, biol. Nauki 20 (7): 102-108, 1977. [In R.]

8002 - UDOVENKO, G.V., GONCHAROVA, Ė.A., KARMANOV, V.G., RADCHENKO, S.S.: Kineticheskie issledovaniya zavisimosti skorosti vodnogo potoka v rastenii ot vneshnikh uslovil. [Kinetic studies of dependence of water transport rate in plants on external conditions.] - Fiziol. Rast. 27: 132-139, 1980. [In R, ab: E.]

8003 - UDOVENKO, G.V., KOZHUSHKO, N.N.: Informativnost' nekotorykh fiziologo-biokhimicheskikh parametrov v svyazi s ustolchivost'yu sortov pshenitsy k zasukhe. [Some physiological and biochemical parameters as an information source for studying the wheat drought resistance.] - Sel'skokhoz. Biol. 15: 359-365, 1980. [In R, ab: E.]

8004 - ÚLEHLA, J.: Stanovení vodního sytostního deficitu dosycováním terčů. Objektivní určení opravy na růst. [Estimation of water saturation deficit by the disc method. Objective evaluation of correction for growth.] - In: Dny Rostlinné Fyziologie II. Pp. 397-401. Vysoká Škola Zemědělská, Brno 1980. [In Czech, ab: E,R.]

8005 - ÚLEHLA, J., ZICHOVÁ, L.: Vliv závlahy na vodní provoz a růst listů cukrovky. [The effect of irrigation on the water regime and growth of leaves in sugar-beet.] - Rost. Výroba (Praha) 26: 509-516, 1980. [In Czech, ab: E,G,R.]

*8006 - UNGAR, I.A.: Seed dimorphism in Salicornia europaea L. - Bot. Gaz. 140: 102-108, 1979.

*8007 - UNGAR, I.A., BENNER, D.K., McGRAW, D.C.: The distribution and growth of Salicornia europaea on an inland salt pan. - Ecology 60: 329-336, 1979.

*8008 – UNGAR, I.A., BOUCAUD, J.: Action des fortes teneurs en NaCl sur l'évolution des cytokinines au cours de la germination d'un halophyte: le *Suaeda maritima* (L.) Dum. var. *macrocarpa* Moq. – C. R. Acad. Sci. Paris, Sér. D 281: 1239-1242, 1975.

8009 – URIU, K., CARLSON, R.M., HENDERSON, D.W., SCHULBACH, H., ALDRICH, T.M.: Potassium fertilization of prune trees under drip irrigation. – J. Amer. Soc. hort. Sci. 105: 508-510, 1980.

8010 – VÁCLAVÍK, J.: Photosynthesis, transpiration and stomatal conductance in *Zea mays* leaves as affected by their irradiance at normal and inverse position. – Photosynthetica 14: 482-488, 1980.

8011 – VÁCLAVÍK, J.: Vliv snižování difusní vodivosti průduchů při ozařování adaxiálního a abaxiálního povrchu listů kukuřice na vztah mezi fotosyntetickým příjmem CO_2 a transpirací. [The influence of the decreasing stomatal conductance during irradiation of adaxial and abaxial leaf surfaces of *Zea mays* on the relationship between photosynthetic CO_2 uptake and transpiration rates.] – In: Dny Rostlinné Fyziologie II. Pp. 416-419. Vysoká Škola Zemědělská, Brno 1980. [In Czech, ab: E,R.]

*8012 – Van der PLOEG, R.R., BEESE, F., BENECKE, P.: Simulationsmodelle von Wald-Ökosystemen: Wasser. – Verh. Ges. Ökol. 1976: 29-41, 1976.

*8013 – Van der WAL, A.F., SMEITING, H., MAAN, G.C.: An ecophysiological approach to crop losses exemplified in the system wheat, leaf rust and glume blotch. III. Effects of soil-water potential on development, growth, transpiration, symptoms, and spore production of leaf rust-infected wheat. – Neth. J. Plant Pathol. 81: 1-13, 1975.

8014 – VAN DIE, J., WILLEMSE, N.C.M.: The supply of water and solutes by phloem and xylem to growing fruits of *Yucca flaccida* Haw. – Ber. Deut. bot. Ges. 93: 327-337, 1980.

8015 – Van KEULEN, H., Van LAAR, H.H., LOUWERSE, W., GOUDRIAAN, J.: Physiological aspects of increased CO_2 concentration. – Experientia 36: 786-792, 1980.

8016 – VAN VOLKENBURGH, E., CLELAND, R.E.: Proton excretion and cell expansion in bean leaves. – Planta 148: 273-278, 1980.

8017 – Van WIJK, A.L.M.: Soil water conditions and playability of grass sportsfields. I. Influence of soil physical properties of top layer and subsoil. II. Influence of tile drainage and sandy drainage layer. – Z. Vegetationstech. 3 (1): 7-22, 1980.

*8018 – VARTHA, E.W.: Growth of 'Grasslands Nui' ryegrass after water and high temperature stress. – N. Zeal. J. agr. Res. 22: 83-87, 1979.

8019 – VASILIU, M.: Contribuţii la stabilirea regimului de irigare la grîul de toamnă în zona Bărăganului de nord. [New results in establishing the irrigation regime for winter wheat in the north Baragan zone.] – An. Inst. Cercetări Pentru Cereale Plante tehnice Fundulea 45: 315-325, 1980. [In Roum, ab: E,R.]

8020 – VENENI, M., AXAMÍT, M.: Využitie strniskových meziplodín na zelený krm pri závlahe. [The use of stubble catch crops grown for green fodder under irrigation.] – Rost. Výroba (Praha) 26: 651-660, 1980. [In Slov, ab: E,G,R.]

8021 – VENKATARAYAPPA, T., TSUJITA, M.J., MURR, D.P.: Influence of cobaltous ion (Co^{2+}) on the postharvest behavior of 'Samantha' roses. – J. Amer. Soc. hort. Sci. 105: 148-151, 1980.

*8022 - VERETENNIKOV, A.V., KONOVALOV, V.N.: Deĭstvie mineral'nykh udobreniĭ i osushe-
niya na potentsial'nyĭ fotosintez *Picea abies* Karst. *(Pinaceae)*. [The effect
of mineral fertilizers and drainage on potential photosynthesis of *Picea abies*
Karst. *(Pinaceae)*.] - Bot. Zh. 64: 80-84, 1979. [In R, ab: E.]

*8023 - VESELOVSKIĬ, V.A., VESELOVA, T.V.: Issledovanie pervichnykh mekhanizmov ustoĭ-
chivosti i adaptatsii resteniĭ k neblagopriyatnym faktoram sredy (biofiziches-
kiĭ podkhod). [Studying of primary mechanisms of resistance and adaptation of
plants to unfavourable environmental conditions (biophysical approach).] - In:
Vsesoyuznoe Soveshchanie: Fiziologo-Biokhimicheskie i Ekologicheskie Aspekty
Ustoĭchivosti Resteniĭ k Neblagopriyatnym Faktoram Vneshneĭ Sredy. Pp. 27-28.
Sibirskiĭ Institut Fiziologii i Biokhimii Resteniĭ, Irkutsk 1976. [In R.]

*8024 - VIANE, R., VAN COTTHEM, W.: Spore morphology and stomatal types in the fern
genera *Asplenium ceterach* and *Phyllitis* *(Aspleniaceae)*. - Acta bot. Neerl. 27:
435, 1978.

*8025 - VIJAYA KUMAR, C.S.K., RAO, A.S.: Transpiration in *Alternaria* infected wheat
plants. - Ind. Phytopathol. 32: 448-450, 1979.

*8026 - VIJAYA KUMAR, C.S.K., RAO, A.S.: Inoculum potential, disease development and
penetration of host by *Alternaria triticina*. Incitant of leaf blight of wheat.
- Proc. Ind. Acad. Sci. 88 B: 359-365, 1979.

8027 - VINOPAL, R.T., WARTELL, S.A., KOLOWSKY, K.S.: β-galactosidase from osmotic
remedial lactose utilization mutants of *E. coli*. - In: RAINS, D.W., VALENTINE,
R.C., HOLLAENDER, A. (ed.): Genetic Engineering of Osmoregulation. Impact on
Plant Productivity for Food, Chemicals and Energy. Pp. 59-72. Plenum Press,
New York - London 1980.

8028 - VIRZO de SANTO, A., ALFANI, A., GRECO, L., FIORETTO, A.: Environmental influences on CAM activity of *Cissus quadrangularis*. - J. exp. Bot. 31: 75-82, 1980.

*8029 - VOLKOVA, R.I.: Vliyanie khlorkholinkhlorida na rost i produktivnost' karto-
felya v zavisimosti ot sortovykh osobennosteĭ i usloviĭ vodosnabzheniya. [In-
fluence of chlorcholinchloride on growth and productivity of potato plants as
affected by variety properties and water supply.] - In: Ekologo-Fiziologicheskie Mekhanizmy Ustoĭchivosti Resteniĭ k Deĭstviyu Ekstremal'nykh Temperatur. Pp. 128-137. Institut Biologii Karel'skogo Filiala Akademii Nauk SSSR,
Petrozavodsk 1978. [In R.]

*8030 - VOLKOVA, R.I.: Kharakter postupleniya i raspredeleniya CCC po organam resteniĭ
kartofelya i soderzhanie ego ostatochnykh kolichestv v klubnyakh. [Character
of CCC uptake and distribution in potato plant organs and its rest content in
tubers.] - In: Ekologo-Fiziologicheskie Mekhanizmy Ustoĭchivosti Resteniĭ k
Deĭstviyu Ekstremal'nykh Temperatur. Pp. 137-145. Institut Biologii Karel'skogo Filiala Akademii Nauk SSSR, Petrozavodsk 1978. [In R.]

*8031 - VOLODARSKIĬ, N.I., BYSTRYKH, E.E.: Funktsional'naya aktivnost' fotosinteti-
cheskogo apparata pri narushenii vodnogo rezhima podsolnechnika. [Functional
activity of the photosynthetic apparatus in the sunflower during disturbance
of the water state.] - Fiziol. Rast. 23: 497-501, 1976. [In R, ab: E.]

8032 - VOLODARSKY, N.I.: Ontogenetic changes of primary photosynthesis reactions in
annual plants. - Acta Agr. Scand. 30: 219-224, 1980.

8033 - VOROBEĬKOV, G.A.: Rol' pobegov kushcheniya yachmenya i pshenitsy v ustoĭchivosti resteniĭ k izbytku vlagi. [The role of barley and wheat tillers in resistance of plants to flooding.] - Sel'skokhoz. Biol. 15: 461-463, 1980. [In
R.]

8034 - WAGENET, R.J., CAMPBELL, W.F., BAMATRAF, A.M., TURNER, D.L.: Salinity, irrigation frequency, and fertilization effects on barley growth. - Agron. J. 72:
969-974, 1980.

*8035 - WAGNER, T.L., GAGNE, J.A., DORAISWAMY, P.C., COULSON, R.N., BROWN, K.W.: Development time and mortality of *Dendroctonus frontalis* in relation to changes in tree moisture and xylem water potential. - Environ. Entomol. 8: 1129-1138, 1979.

*8036 - WALIA, R.S., SINGH, R., SINGH, Y.: Yield and water use of dryland wheat as affected by N application and soil type in a maize-wheat sequence. - Ind. J. Ecol. 6: 60-67, 1979.

8037 - WALLACE, L.L., DUNN, E.L.: Comparative photosynthesis of three gap phase succesional tree species. - Oecologia 45: 331-340, 1980.

8038 - WALTON, D.C.: Biochemistry and physiology of abscisic acid. - Annu. Rev. Plant Physiol. 31: 453-489, 1980.

8039 - WANG, J.R., SCHMUGGE, T.J.: An empirical model for the complex dielectric permittivity of soils as a function of water content. - IEEE Trans. Geosci. Remote Sensing GE-18 (4): 288-295, 1980.

8040 - WANNER, H., SOEROHALDOKO, S.: Transpiration types in montane rain forest. - Ber. Schweiz. bot. Ges. 89: 193-210, 1980.

8041 - WARDLAW, I.F.: Translocation and source-sink relationships. - In: CARLSON, P. S. (ed.): The Biology of Crop Productivity. Pp. 297-333. Academic Press, New York - London - Toronto - Sydney - San Francisco 1980.

*8042 - WAREMBOURG, F.R., PAUL, E.A.: Seasonal transfers of assimilated ^{14}C in grassland: plant production and turnover, soil and plant respiration. - Soil Biol. Biochem. 9: 295-301, 1977.

8043 - WARING, R.H., WHITEHEAD, D., JARVIS, P.G.: Comparison of an isotopic method and the Penman-Monteith equation for estimating transpiration from Scots pine. - Can. J. Forest Res. 10: 555-558, 1980.

8044 - WARRICK, A.W., LOMEN, D.O., AMOOZEGAR-FARD, A.: Linearized moisture flow with root extraction for three dimensional, steady conditions. - Soil Sci. Soc. Amer. J. 44: 911-914, 1980.

8045 - WATERS, L.,Jr., GRAHAM, P.H., BREEN, P.J., MACK, H.J., ROSAS, J.C.: The effect of rice-hull mulch on growth, carbohydrate content, and nitrogen fixation in *Phaseolus vulgaris* L. - HortScience 15: 138-140, 1980.

8046 - WATT, T.A., HAGGAR, R.J.: The effect of height of water table on the growth of *Holcus lanatus* with reference to *Lolium perenne*. - J. appl. Ecol. 17: 423-430, 1980.

8047 - WEBB, R.O., CULVER, E.D., LAULAINEN, N.S.: Calibration of special water sensitive paper including droplet impaction at oblique angles. - Atmos. Environ. 14: 385-389, 1980.

*8048 - WEGMANN, K.: Biochemische Anpassung von *Dunaliella* an wechselnde Salinität und Temperatur. - Ber. Deut. bot. Ges. 92: 43-54, 1979.

*8049 - WEIGEL, H.-J., JÄGER, H.-J.: Changes in proline concentration of the lichen *Pseudevernia furfuracea* during drought stress. - Phyton 19: 163-167, 1979.

*8050 - WEIHE, K., von: Morphologische und okologische Grundlagen der Vorlandsicherung durch *Puccinellia maritima Gramineae* . - Helgoländer wiss. Meeresunters. 32: 239-254, 1979.

8051 - WEILAND, R.T., STUTTE, C.A.: Concomitant determination of folar nitrogen loss, net carbon dioxide uptake, and transpiration. - Plant Physiol. 65: 403-406, 1980.

*8052 - WEILAND, R.T., STUTTE, C.A., TALBERT, R.E.: Foliar nitrogen loss and CO_2 equilibrium as influenced by three soybean (*Glycine max*) postemergence herbicides. - Weed Sci. 27: 545-548, 1979.

8053 - WELLS, J.A., NUGENT, P.E.: Effect of high soil moisture on quality of muskmelon. - HortScience 15: 258-259, 1980.

8054 - WENKERT, W.: Measurement of tissue osmotic pressure. - Plant Physiol. 65: 614-617, 1980.

8055 - WEST, D.W., MERRIGAN, I.F., TAYLOR, J.A., COLLINS, G.M.: Growth of ornamental plants irrigated with nutrient or polyethylene glycol solutions of different osmotic potentials. - Plant Soil 56: 99-111, 1980.

8056 - WEST, D.W., TAYLOR, J.A.: The effect of temperature on salt uptake by tomato plants with diurnal and nocturnal waterlogging of salinized rootzones. - Plant Soil 56: 113-121, 1980.

8057 - WEST, D.W., TAYLOR, J.A.: The response of *Phaseolus vulgaris* L. to root-zone anaerobiosis, waterlogging and high sodium chloride. - Ann. Bot. 46: 51-60, 1980.

8058 - WEST, L.D., DAWSON, J.H., APPLEBY, A.P.: Factors influencing barnyagrass (*Echinochloa crus-galli*) control with diclofop. - Weed. Sci. 28: 366-371, 1980.

*8059 - WEST, P.W.: Date of onset of regrowth dieback and its relation to summer drought in eucalypt forest of southern Tasmania. - Ann. appl. Biol. 93: 337-350, 1979.

8060 - WEYERS, J.D.B., HILLMAN, J.R.: Effects of abscisic acid on $^{86}Rb^+$ fluxes in *Commelina communis* L. leaf epidermis. - J. exp. Bot. 31: 711-720, 1980.

8061 - WHITEHEAD, D.: Assessment of water status in trees from measurements of stomatal conductance and water potential. - N. Zeal J. Forest. Sci. 10: 159-165, 1980.

8062 - WIEBE, H.H.: Morphological adaptations to water stress. - In: TURNER, N.C., KRAMER, P.J. (ed.): Adaptation of Plants to Water and High Temperature Stress. Pp. 439-443. John Wiley & Sons, New York - Chichester - Brisbane - Toronto 1980.

8063 - WIENCKE, C., LÄUCHLI, A.: Growth, cell volume, and fine structure of *Porphyra umbilicalis* in relation to osmotic tolerance. - Planta 150: 303-311, 1980.

8064 - WIENEKE, J., LÄUCHLI, A.: Effects of salt stress on distribution of Na^+ and some other cations in two soybean varieties differing in salt tolerance. - Z. Pflanzenernähr. Bodenk. 143: 55-67, 1980.

*8065 - WIGHT, J.R., BLACK, A.L.: Range fertilization: Plant response and water use. - J. Range Manage. 32: 345-349, 1979.

8066 - WILD, A., WOLF, G.: The effect of different light intensities on the frequency and size of stomata, the size of cells, the number, size and chlorophyll content of chloroplasts in the mesophyll and the guard cells during the ontogeny of primary leaves of *Sinapis alba*. - Z. Pflanzenphysiol. 97: 325-342, 1980.

*8067 - WILKINSON, J.F., BEARD, J.B.: Anatomical responses of 'Merion' Kentucky bluegrass and 'Pennlawn' red fescue at reduced light intensities. - Crop Sci. 15: 189-194, 1975.

8068 - WILLERT, D.J., von, BRINCKMANN, E., SCHEITLER, B., SCHULZE, E.-D., THOMAS, D. A., TREICHEL, S.: Ökophysiologische Untersuchungen an Pflanzen der Namib-Wüste. - Naturwissenschaften 67: 21-28, 1980.

8069 - **WILLET, I.R., HIGGINS, M.L.:** Phosphate sorption and extractable iron in soils during irrigated rice-upland crop rotations. - Aust. J. exp. Agr. anim. Husb. 20: 346-353, 1980.

8070 - **WILLIAMS, R.D.:** Moisture stress and hydratation-dehydratation effects on hemp sesbania (*Sesbania exaltata*) seed germination. - Weed Sci. 28: 487-492, 1980.

8071 - **WILLMER, C.M.:** Some characteristics of phosphoenolpyruvate carboxylase activity from leaf epidermal tissue in relation to stomatal functioning. - New Phytol. 84: 593-602, 1980.

8072 - **WILSON, D.R., VAN BAVEL, C.H.M., McCREE, D.R.:** Carbon balance of water-deficient grain sorghum plants. - Crop Sci. 20: 153-159, 1980.

8073 - **WILSON, J.R., LUDLOW, M.M., FISHER, M.J., SCHULZE, E.-D.:** Adaptation to water stress of the leaf water relations of four tropical forage species. - Aust. J. Plant Physiol. 7: 207-220, 1980.

8074 - **WILSON, S.J., JONES, K.M.:** Responses of blackcurrant bushes to post-harvest moisture stress. - Scientia Hort. 12: 307-312, 1980.

8075 - **WINNER, W.E., MOONEY, H.A.:** Responses of hawaiian plants to volcanic sulfur dioxide: Stomatal behavior and foliar injury. - Science 210: 789-791, 1980.

8076 - **WINNER, W.E., MOONEY, H.A.:** Ecology of SO_2 resistance: I. Effects of fumigations on gas exchange of deciduous and evergreen shrubs. - Oecologia 44: 290-295, 1980.

8077 - **WINNER, W.E., MOONEY, H.A.:** Ecology of SO_2 resistance: II. Photosynthetic changes of shrubs in relation to SO_2 absorption and stomatal behavior. - Oecologia 44: 296-302, 1980.

8078 - **WINNER, W.E., MOONEY, H.A.:** Ecology of SO_2 resistance: III. Metabolic changes of C_3 and C_4 *Atriplex* species due to SO_2 fumigations. - Oecologia 46: 49-54, 1980.

8079 - **WINTER, K.:** Carbon dioxide and water vapor exchange in the Crassulacean acid metabolism plant *Kalanchoë pinnâta* during a prolonged light period. Metabolic and stomatal control of carbon metabolism. - Plant Physiol. 66: 917-921, 1980.

*8080 - **WINTER, K., LÜTTGE, U.:** C_3-Photosynthese und Crassulaceen-Säurestoffwechsel bei *Mesembryanthemum crystallinum* L. - Ber. Deut. bot. Ges. 92: 117-132, 1979.

8081 - **WINTER, S.R.:** Suitability of sugarbeets for limited irrigation in a semi-arid climate. - Agron. J. 72: 118-123, 1980.

*8082 - **WISIOL, K.:** Clipping of water-stressed blue grama affects proline accumulation and productivity. - J. Range Manage. 32: 194-195, 1979.

8083 - **WITTENBACH, V.A., ACKERSON, R.C., GIAQUINTA, R.T., HEBERT, R.R.:** Changes in photosynthesis, ribulose bisphosphate carboxylase, proteolytic activity, and ultrastructure of soybean leaves during senescence. - Crop Sci. 20: 225-231, 1980.

8084 - **WITTWER, S.H.:** The shape of things to come. - In: CARLSON, P.S. (ed.): The Biology of Crop Productivity. Pp. 413-459. Academic Press, New York - London - Toronto - Sydney - San Francisco 1980.

8085 - **WOLFE, E.C., SOUTHWOOD, O.R.:** Plant productivity and persistence in mixed pastures containing lucerne at a range densities with subterranean clover or phalaris. - Aust. J. exp. Agr. anim. Husb. 20: 189-196, 1980.

*8086 - WONG, B.S., MILLER, D.M., YOPP, J.H.: Proton pulsed NMR study on the cell
        constituents of *Aphanothece halophytica*, a blue-green alga. - In: AGRIS, P.F.,
        LOEPPKY, R.N., SYKES, B.D. (ed.): Biomolecular Structure and Function. Pp. 239
        -245. Academic Press, New York - San Francisco - London 1978.

8087 - WONG, C.C., WILSON, J.R.: Effects of shading on the growth and nitrogen con-
        tent of green panic and Siratro in pure and mixed swards defoliated at two
        frequencies. - Aust. J. agr. Res. 31: 269-285, 1980.

8088 - WRIGHT, J.P., FISHER, D.B.: Direct measurement of sieve tube turgor pressure
        using severed aphid stylets. - Plant Physiol. 65: 1133-1135, 1980.

8089 - WRIGHT, S.T.C.: The effect of plant growth regulator treatments on the levels
        of ethylene emanating from excised turgid and wilted wheat leaves. - Planta
        148: 381-388, 1980.

8090 - WU, Y.-D., HE, J.: [The bound water content of Brazil rubber (*Hevea brasilien-
        sis*) as an index of cold resistance.] - Acta phytophysiol. Sinica 6: 107-114,
        1980. [In Chin, ab: E.]

8091 - WYN JONES, R.G.: An assessment of quaternary ammonium and related compounds
        as osmotic effectors in crop plants. - In: RAINS, D.W., VALENTINE, R.C.,
        HOLLAENDER, A. (ed.): Genetic Enginnering of Osmoregulation. Impact on Plant
        Productivity for Food, Chemicals, and Energy. Pp. 155-170. Plenum Press, New
        York - London 1980.

*8092 - WYRILL, J.B.,III., BURNSIDE, O.C.: Absorption, translocation, and metabolism
        of 2,4-D and glyphosate in common milkweed and hemp dogbane. - Weed Sci. 24:
        557-566, 1976.

8093 - XILOYANNIS, C., URIU, K., MARTIN, G.C.: Seasonal and diurnal variations in
        abscisic acid, water potential, and diffusive resistance in leaves from irri-
        gated and non-irrigated peach trees. - J. Amer. Soc. hort. Sci. 105: 412-415,
        1980.

88094 - YABLONSKIĬ, E.A.: Vzaimosvyaz' vodnogo rezhima generativnykh pochek i odno-
        letnikh pobegov abrikosa v period zimnego razvitiya. [Relationship between
        water regime in generative buds and in annual apricot shoots during winter
        development.] - Fiziol. Rast. 27: 828-834, 1980. [In R, ab : E.]

*8095 - YADAV, T.D., PANT, N.C.: Moisture content - relative humidity relationship of
        legume seeds. - Seed Res. 7: 11-17, 1979.

*8096 - YAKOVLEV, B.V., YAKOVLEVA, V.F., ALESHIN, E.P.: Primenenie pozdnoĭ vnekornevoĭ
        podkormki fosforom i azotom v sochetanii s 2,4-DA dlya uskoreniya sozrevaniya
        zerna i povysheniya urozhaya risa. [Utilization of late foliar phosphorus and
        nitrogen nutrition together with 2,4-DA for accelerating grain ripening and
        yield increase.] - Byull. nauch.-tekh. Informatsii vses. nauch.-issled. Inst.
        Risa 15: 30-36, 1975.[In R, ab: E.]

8097 - YAKOVLEVA, G.A., MOLOTKOVSKIĬ, Yu.G.: Vliyanie *p*-khlormerkuribenzoata na
        ionnuyu pronitsaemost' membran tilakoidov. [Effect of *p*-chloromercuribenzoate
        on ion permeability of thylakoid membranes.] - Fiziol. Rast. 27: 710-721,
        1980. [In R, ab: E.]

8098 - YAMAMOTO, T., WATANABE, S.: [Measurements of sap velocities in stems,
        peduncles and petioles of pear trees by a heat pulse method.] - J. Jap. Soc.
        hort. Sci. 311-325, 1980. [In Jap, ab: E.]

8099 - YAMANAKA, S., OZAKI, M., KATO, S.: Some observations on angular leaf spot of
        cucumber with a scanning electron microscope. - Tohoku J. agr. Res. 30: 135-
        141, 1980.

8100 - YANG, S.F., ADAMS, D.O., LIZADA, C., YU, Y., BRADFORD, K.J., CAMERON, A.C.,
HOFFMAN, N.E.: Mechanism and regulation of ethylene biosynthesis. - In:
SKOOG, F. (ed.): Plant Growth Substances 1979. Pp. 219-229. Springer-Verlag,
Berlin - Heidelberg - New York 1980.

8101 - YEGAPPAN, T.M., PATON, D.M., GATES, C.T., MÜLLER, W.J.: Water stress in sunflower (*Helianthus annuus* L.): I. Effect on plant development. - Ann. Bot. 46:
61-70, 1980.

8102 - YEO, A.R., FLOWERS, T.J.: Salt tolerance in the halophyte *Suaeda maritima* L.
Dum.: Evaluation of the effect of salinity upon growth. - J. exp. Bot. 31:
1171-1183, 1980.

*8103 - YOSHIDA, S.: Rice; - In: ALVIM, P. de T., KOZLOWSKI, T.T. (ed.): Ecophysiology
of Tropical Crops. Pp. 57-87. Academic Press, New York - San Francisco - London 1977.

*8104 - YOSHIDA, S., CORONEL, V.: Nitrogen nutrition, leaf resistance, and leaf photo-
synthetic rate of the rice plant. - Soil Sci. Plant Nutr. 22: 207-211, 1976.

*8105 - YOSHIDA, T.: [On the stomatal frequency in barley. II. Intra-plant variation,
varietal difference and heritability of stomatal frequency.] - Jap. J. Breed.
27: 91-97, 1977. [In Jap, ab: E.]

8106 - YOUNG, E., HOUSER, J.: Influence of Siberian C rootstock on peach bloom delay,
water potential, and pollen meiosis. - J. Amer. Soc. hort. Sci. 105: 242-245,
1980.

*8107 - YOUNG, R.H., GARNSEY, S.M., HORANIC, G.: A device for infusing liquids into
the outher xylem vessels of citrus trees. - Plant Dis. Rep. 63: 713-715, 1979.

8108 - YOUNG, R.H., WUTSCHER, H.K., ALBRIGO, L.G.: Relationships between water translocation and zinc accumulation in citrus trees with and without blight. - J.
Amer. Soc. hort. Sci. 105: 444-447, 1980.

*8109 - YOUNIS, H.M., BOYER, J.S., GOVINDJEE: Magnesium binding to chloroplast coup-
ling factor correlated with effects of low water potential. - Plant Physiol.
63 (Suppl.): 40: 1979.

*8110 - YOUSIF, H.J., BARDEN, J.A.: Response of the apple rootstock MM 104 to six soil
mixes and three water levels. - Mesopotamia J. Agr. 14: 131-143, 1979.

*8111 - YUKHNO, G.Ya., KULIK, I.I.: Vlagoobespechennost' i produktivnost' lyutserny v
zavisimosti ot vida pokrovnoi kul'tury, sposobov i srokov poseva. [Water supply and productivity of alfalfa depending upon cover crop species, methods and
time of sowing.] - Byull. vses. nauch.-issled. Inst. Kukuruzy 1975 (2/38):
59-62, 1975. [In R, ab: E.]

*8112 - ZACHAR, D.: K problematike bilancie vody a energie poľnohospodárskych a les-
ných ekosystémov.[On the problem of the water and energy balance of field and
forest ecosystems.] - In: Bilancia Energie a Vody v Poľných a Lesných Ekosystémoch. Pp. 7-15. Vysoká Škola Poľnohospodárská, Nitra 1979. [In Slov.]

8113 - ZEEVAART, J.A.D.: Changes in the levels of abscisic acid and its metabolites
in excised leaf blades of *Xanthium strumarium* during and after water stress. -
Plant Physiol. 66: 672-678, 1980.

*8114 - ZEHR, D.R.: Phenology of selected bryophytes in southern Illinois. - Bryo-
logist 82: 29-36, 1979.

8115 - ZEIGER, E.: The blue light response of stomata and the green vacuolar fluorescence of guard cells. - In: SENGER, H. (ed.): The Blue Light Syndrome. Pp.
629-636. Springer-Verlag, Berlin - Heidelberg - New York 1980.

8116 - ZEIGER, E., ARMOND, P., MELIS, A.: Fluorescence properties of guard cell chloroplasts. Evidence for linear electron transport and light-harvested pigments of photosystems I and II. - Plant Physiol. 67: 17-20, 1980.

8117 - ZELEŇÁKOVÁ, E., POLEK, B.: Matematicko-štatistické vyhodnotenie vplyvu klimatických faktorov na obsah chlorofylu v listoch marhule (Prunus armeniaca L.) [Mathematical-statistical evaluation of the influence of climatic factors on chlorophyll content in leaves of the apricot-tree (Prunus armeniaca L.).] - Biológia (Bratislava) 35: 51-58, 1980. [In Slov, ab: E,R.]

8118 - ZEMÁNEK, M.: Genotypové rozdíly v efektivnosti využívání vody na tvorbu sušiny u jarního ječmene. [Genotypic differences in the efficiency of water utilization for dry matter production in spring barley.] - Rost. Výroba (Praha) 26: 517-531, 1980. [In Czech, ab: E,G,R.]

8119 - ZEMÁNEK, M.: Reakce genotypů jarního ječmene na rozdílné zásobení vodou. [The response of spring barley genotypes to different water supply.] - Rost. Výroba (Praha) 26: 1291-1305, 1980. [In Czech, ab: E,G,R.]

*8120 - ZENTMYER, G.A.: Effect of physical factors, host resistance and fungicides on root infection at the soil-root interface. - In: HARLEY, J.L., SCOTT-RUSSELL, R. (ed.): The Soil-Root Interface. Pp. 315-328. Academic Press, London - New York - San Francisco 1979.

*8121 - ZHELEV, R.: Razrastvane na korenovata sistema na tsarevitsata pri napoyavane s reguliruemi ustroĭstva. [The growth of maize roots following irrigation with regulating devices.] - Rasteniev. Nauki 16 (9-10): 60-64, 1979. [In Bulg, ab: E,R.]

*8122 - ZHOLKEVICH, V.N., SINITSYNA, Z.A., PEĬSAKHZON, B.I., ABUTALYBOV, V.F., D'YACHENKO, I.V.: O prirode kornevogo davleniya. [The nature of root pressure.] - Fiziol. Rast. 26: 978-993, 1979. [In R, ab: E.]

*8123 - ZIMMERMANN, U., BECKERS, F., STEUDLE, E.: Turgor sensing in plant cells by electro-mechanical properties of the membrane. - In: THELLIER, M., MONNIER, A., DEMARTY, M., DAINTY, J. (ed.): Transmembrane Ionic Exchanges in Plants. Pp. 155-165. Centre National de la Recherche Scientifique, Paris 1977.

8124 - ZIMMERMANN, U., HÜSKEN, D.: Turgor pressure and cell volume relaxation in Halicystis parvula. - J. Membrane Biol. 56: 55-64, 1980.

8125 - ZIMMERMANN, U., HÜSKEN, D., SCHULZE, E.-D.: Direct turgor pressure measurements in individual leaf cells of Tradescantia virginiana. - Planta 149: 445-453, 1980.

8126 - ZIMMERMANN, U., STEUDLE, E.: Fundamental water relations parameters. - In: SPANSWICK, R.M., LUCAS, W.J., DAINTY, J. (ed.): Plant Membrane Transport: Current Conceptual Issues. Pp. 113-127. Elsevier / North-Holland Biomedical Press, Amsterdam - New York 1980.

8127 - ZOBEL, D.B., LIU, V.T.: Effects of environment, seedling age, and seed source on leaf resistance of three species of Chamaecyparis and Tsuga chinensis. - Oecologia 46: 412-419, 1980.

8128 - ZOBEL, D.B., LIU, V.T.: Leaf-conductane patterns of seven palms in a common environment. - Bot. Gaz. 141: 283-289, 1980.

AUTHORS' INDEX

Authors' names are presented in the form in which they appear in the respective
publication. The names from papers published in Cyrillic character are transcribed
as shown in Instructions for Use. Alternative spelling and form of the name of the
same author are usually cross-indexed.

A

AASE, J.K. 6606
ABRAMOV, V.L. 7332
ABROL, B.K. 6607
ABROL, I.P. 6610
ABRUÑA, F. 6608
ABUTALYBOV, V.F. 8122
ACEVEDO, E. 6609
ACHARYA, C.L. 6610
ACKERSON, R.C. 6611, 6612, 8083
ACKLEY, W.B. 6613
ADAMS, D.E. 6614
ADAMS, D.O. 8100
ADDICOT, F.F. 6615
ADJEI, G.B. 6987
AFANAS'EV, V.P. 7411
AGABBIO, M. 6616
AGATA, W. 7942
AGBOOLA, A.A. 7000
AGGARWAL, R.K. 6617, 7099
AGUILAR, G.J.R. 6845
AHMED, A.M. 6618, 7144
AHMED, S. 6619
AHMED, Z.U. 7405
AHO, N. 6620, 6621
AKALEHIYWOT, T. 6622
AKITA, S. 6623, 6624
AKOPYANTS, N.S. 7510
ALBERS, D.J. 6625
ALBERT, R. 6626, 7301
ALBERTE, R.S. 6866
ALBINET, E. 6627
ALBREGTS, E.E. 6628
ALBRIGO, L.G. 6632, 7028, 8108
ALDRICH, T.M. 8009
ALESHIN, E.P. 8096
ALESSI, J. 6629
ALFANI, A. 8028
ALI, H.C. 6630
AL-ITHAWI, B. 6631
ALLEN, J.J. 6632
ALLEN, R.R. 6633
ALONI, B. 6634
ALPERT, P. 6635
ALSTON, A.M. 6636
ALVAREZ, D. 7148
ALVIM, P.de T. 6637, 6638

ALZUBAIDI, A.H. 6639
AMOOZEGAR-FARD, A. 8044
ANDALES, S.C. 6640
ANDEREGG, J.C. 7408
ANDERSEN, A.S. 7675, 7676, 7687
ANDERSEN, R.L. 6810
ANDERSON, J.E. 6641, 7413
ANDERSON, R.C. 6614
ANDERSON, W.K. 6642, 6643
ANDONOVA, P. 7320
ANDRÉ, M. 7425
ANDREI, R. 7271
ANGUILLESI, M.C. 6644
ANGUS, J.F. 6645
ANIKEENKO, A.P. 6929
ANNAMMA, Y. 7648
ANNAN, A.P. 7980
ANTIPOV, N.I. 6646
ANTLFINGER, A.E. 6647
APARICIO-TEJO, P.M. 6648, 6649
APEL, P. 6650
APPELBAUM, S. 6651
APPLEBY, A.P. 6846, 8058
AQUINO, A.R.L., de see De AQUINO, A.R.L.
ARDITTI, J. 6652
AREKAL, G.D. 7513
ARIAS, I. 7453
ARMOND, P. 8116 see also
ARMOND, P.A. 6731, 7076, 7673
ARTYKOV, K. 6653
ASADA, S. 6654
ASBELL, C.W. 7993
ASHCROFT, G.L. 7111
ASHCROFT, W.J. 6655
ASPINALL, D. 6656
ASTON, A.R. 6657
ATKINS, C.A. 7603
ATTIWILL, P.M. 6658
AUGSPURGER, C.K. 6659
AUSSENAC, G. 6660, 6661
AVARVAREI, I. 7963
AVRON, M. 6709, 6710
AXAMIT, M. 8020
AXLEY, J.H. 7611
AYRES, P.G. 6662, 6663
AZUARA, P. 7756, 7757

B

BABBER, S. 7823
BACHELARD, E.P. 6664, 6665
BACON, G.J. 6664, 6665
BADGER, M.R. 6731
BAGNALL, D. 6666
BAIER, W. 6962
BAILISS, K.W. 6667
BAKER, D.N. 6668
BAKER, E.A. 6669, 6793
BAKER, N.R. 6938, 6939, 6940
BAKRADZE, N.G. 6670, 6671
BAKSHI, R.K. 7410
BALAAM, L.N. 7072
BALINA, N.V. 6672
BALLA, Yu.I. 6670
BALLARD, T.M. 6828
BAMATRAF, A.M. 8034
BANERJI, M.L. 6673
BANGE, G. 6674
BANGERTH, F. 6988
BAÑOCH, Z. 6675, 6676
BANOV, I. 7311
BANSAL, R.P. 6677
BARASSI, C.A. 6678
BARBER, J. 6679
BARDEN, J.A. 6680, 8110
BARKUS, B. 7994
BARLOW, E.W.R. 6681, 6682, 7798, 7900
BARNES, J.E. 6683
BARRY, G. 7925
BARTA, V. 6922
BARTHOLOMEW, D.P. 6684
BARTLE, G.A. 6819
BARTLETT, B.O. 7831
BARTON, J.R. 6685
BARUAH, K.K. 6686
BAR-YOSEF, B. 6687
BASU, R.N. 6688, 6689
BAUDER, J.W . 6690
BAUER, H. 6691, 7305
BAUER, U. 6691
BAVEL, C.H.M., van see VAN BAVEL, C.H.M.
BAZZAZ, F.A. 6823
BEALE, O.W. 7632
BEARD, J.B. 8067
BEARDSELL, D.V. 6692
BECHSTÄDT, O. 7920
BECKERS, F. 8123
BEDFORD, C.L. 7875
BEDUNAH, D. 6693
BEER, S. 6694
BEESE, F. 6695, 8012
BEGG, J.E. 6696
BÉGUIN, C. 6697
BEGUN, S. 7405
BEHBOUDIAN, M.H. 6698
BEHRNS, G.T. 6931
BEINEKE, W.E. 6699

BELAN, F. 6700
BELAYA, G.A. 6701, 6702
BELFORD, R.K. 6703, 6815, 6816, 6817
BELL, D.H. 7070
BELL, K.L. 6704, 6705
BELLES, W.S. 7887
BELMANS, C. 7011
BELOT, Y. 6706
BELT, G.H. 6707
BEN-AMOTZ, A. 6708, 6709, 6710
BENECKE, P. 6711, 8012
BENECKE, U. 6712, 7936
BENNER, D.K. 8007
BENNIE, A.T.P. 7424
BENOIT, G.R. 6713
BENTLEY, B.L. 6714
BENTRUP, F.W. 7046
BERESFORD, J.D. 7088, 7089
BERGMANN, H. 7295
BERGSRUD, F.G. 7493
BERKALOFF, C. 6715
BERKOWITZ, G.A. 6716
BERRY, J. 6717
BERRY, J.A. 7673, 7802
BESRI, M. 6718
BETHENOD, O. 7490, 7719
BETTS, M.F. 6719
BEVERSDORF, W.D. 6720
BEWLEY, J.D. 6622, 6721
BHARATI, M.P. 6722
BHARDWAJ, R.B.L. 7816
BHASKARAN NAIR, V.K. 7648
BHATI, P.R. 6677
BIBLE, B.B. 6723
BIDDINGTON, N.L. 6724
BIDINGER, F. 6725
BIERE, A.W. 7496
BIGGS, R.H. 7028, 7306, 7961
BIGLOVA, S.G. 7745
BIKHELE, Z.N. 6726
BILLARD, J.P. 6727 see also
BILLARD, J.-P. 6764
BINDEROVÁ, A. 6728
BINGHAM, G.E. 6729
BIRD, I.F. 7972
BIRYUKOV, V.N. 6730
BIRYUKOVA, Z.P. 6730
BISCOE, P.V. 7420, 7421
BJÖRKMAN, O. 6717, 6731, 6732
BJÖRN, L.O. 7933
BLACK, A.L. 8065
BLACK, V.J. 6733
BLACKLOW, W.M. 6734
BLACKWELL, P.S. 6735
BLAKE, J. 6736
BLAKE, T.J. 6737
BLANCHAR, R.W. 6773, 7589
BLANCHET, R. 6738, 6739
BLANCO, A. 7944
BLASZCZAK, W. 7427
BLECKMANN, C.A. 6740
BLEKHMAN, G.I. 6741

STANHILL, G. 7894
STANKOVA, P.G. 7264
STANKOVA, J. 7895
STANLEY, C.D. 7896
STANSELL, J.R. 7897
STARK, J.C. 7898
STATLER, G.D. 7899
STAVAREK, S.J. 7672
STEELE SCOTT, N. 7900
STEGMAN, E.C. 7532
STEHLIK, K. 7901, 7902
STELZER, R. 7903
STEPONKUS, P.L. 6899, 6900, 6901,
7366, 7904, 7905
STEUDLE, E. 7906, 8123, 8126
STEVENINCK, R.F.M., van see
VAN STEVENINCK, R.F.M.
STEWART, C.R. 6746, 7907
STEWART, J.B. 7614
STEWART, R.H. 7265
STIGTER, C.J. 7908
STIGTER, H.C.M., de see
DE STIGTER, H.C.M.
STILES, W. 7231, 7232
STINSON, R.H. 7439
STOBBS, L.W. 7803
STOCK, H.-G. 7909, 7910
STOFFERS, A.L. 7911
STOKER, R. 7912, 7913, 7914
STOLZY, L.H. 7242, 7243, 7442,
7832, 7881
STONE, D.A. 7915, 7916
STONE, E.L. 7045
STONE, J.A. 7424
STONE, J.E. 7917
STONE, J.F. 7918
STONE, L.F. 7919
STÖPEL, W. 7920
STORDEUR, R. 7921
STOUT, D.G. 7922, 7923
STOYANOV, Zh. 7924
ST-PIERRE, C.A. 7219
STRAIN, B.R. 7712, 7851
STREICH, J. 6946
STROGONOV, B.P. 7888, 7889
STRONG, W.M. 7925
STUIVER, B.(C.E.E.) 6986
STUMPE, H. 7926
STUMPF, D.K. 7885
STUSHNOFF, C. 6797
STUTLER, R.K. 7090
STUTTE, C.A. 7927, 8051, 8052
SUAREZ, J.J. 7928
SUBRAMANYAM, P. 7929
SUDNITSYN, I.I. 7930
SUELDO, R.J. 6678
SUGAHARA, K. 7300
SUH, H.W. 7931
SUMAYAO, C.R. 7932
SUNDQVIST, C. 7933
SUNG, F.J.M. 6612
SURANYI, D. 7934

SUTHERLAND, S.M. 7086
SVESHNIKOVA, V.M. 7935
SWAMY, B.G.L. 7513
SWAMY, P.M. 6907, 7397, 7398,
7683, 7684
SWANSON, R.H. 7723, 7936
SYBER, A.Yu. 7937
SYBER, Ya.Kh. 7484
SYKORA, K.V. 7938
SYTNYK, K.M. 7414
SZAREK, S.R. 7939
SZURMAN, N. 6795

T

TAIZ, L. 7940
TAKANO, M. 6654
TAKEBA, G. 7941
TAKEDA, T. 7942
TALBERT, R.E. 8052
TAN, C.S. 7943
TANNER, C.B. 7432
TANZARELLA, O.A. 7944
TARAN, N.Yu. 7564
TARGON, P.G. 7945
TASHIRO, T. 7328
TAUER, C.G. 7268
TAYLOR, A.G. 7691
TAYLOR, F.E. 7946
TAYLOR, H.M. 7424, 7896, 7947,
7948, 7949
TAYLOR, J.A. 7950, 8055, 8056,
8057
TAZAKI, T. 7951
TAZAWA, M. 7132, 7255, 7952
TEARE, I.D. 7852
TEERI, J.A. 6866
TEGLER, B. 7953
TENHUNEN, J.D. 7954, 7955, 7956,
7957
TEN KLOOSTER, W.P. 7091
TEODORESCU, G. 7958, 7959
TERAMURA, A.H. 6911, 7960, 7961
TERAO, H. 7962
TERPUGOV, E.L. 7355
TESKEY, R.O. 6931
TEŞU, C. 7963
TEŞU, V. 7964
THAKUR, P.S. 7965
THAKUR, R.P. 7847
THALOUARN, P. 7966
THANKI, Y.J. 7967
THIMANN, K.V. 7055, 7968
THIMMEGOWDA, S. 7969, 7970
THOMAS, D.A. 8068
THOMAS, D.G. 7368
THOMAS, J.C. 6786
THOMAS, J.D. 6834
THOMAS, J.R. 7971
THOMAS, S. 7972
THOMAS, T.E. 7973

THOMAS, T.H. 6724
THOMSON, R.J. 6703
THORNE, S.W. 6854
THORPE, N. 7974
THORPE, T.A. 6784
THULLEN, R.J. 7975
THURLING, N. 7702, 7703
THURTELL, G.W. 7554
TICHÁ, I. 6833, 7883
TIETZ, D. 6946
TIMMIS, R. 7245
TIMMONS, D.R. 6963
TING, I.P. 7976
TÎNJALĂ, I. 7963
TKACHUK, K.S. 7977
TOMA, L.-D. 7964
TOMAR, V.S. 7978, 7979
TOMES, D.T. 7440
TOMIYA, N. 7563
TOPP, G.C. 7980, 7981
TRAVELLER, D.J. 6829
TREHARNE, K.J. 6996
TREICHEL, S. 7982, 8068
TRLICA, M.J. 6693, 7010
TROUGHT, M.C.T. 7983, 7984
TROUGHTON, A. 7985
TROUGHTON, J.H. 7635
TRUKHAN, É.M. 7861, 7862
TRUKSA, J. 7986
TRUȘCĂ, M. 7215
TSCHAKALOVA, E. 7987
TSUCHIYA, T. 7507
TSUJITA, M.J. 8021
TSUNODA, S. 7988
TUCKER, C.J. 7989
TURK, K.J. 7990, 7991, 7992,
7993
TURKINGTON, C.R. 7029
TURNER, B.J. 6754
TURNER, D.L. 8034
TURNER, D.W. 7994
TURNER, J.L. 6949
TURNER, M.A. 7799
TURNER, N.C. 7233, 7234, 7995,
7996, 7997
TURPIN, D.H. 7973
TUZINSKÝ, L. 7998
TVERGYAK, P.J. 7999
TYLER, E.J. 7211
TYREE, M.T. 6896, 8000
TYSHKEVICH, G.L. 8001

U

UDOVENKO, G.V. 8002, 8003
ÚLEHLA, J. 8004, 8005
UNGAR, I.A. 8006, 8007, 8008
UNGER, H. 7093
UNSWORTH, M.H. 6733
URIU, K. 8009, 8093
USHIJIMA, T. 7951

V

VÁCLAVÍK, J. 7617, 8010, 8011
VAN BAVEL, C.H.M. 8072
VAN COTTHEM, W. 8024
Van der PLOEG, R.R. 6711, 8012
Van der WAL, A.F. 8013
VAN DIE, J. 8014
VAN DIXHOORN, J.J. see DIXHOORN, J.J.,
van
VAN GELDER, H. see GELDER, H., van
VAN KEULEN, H. 7074, 8015
VAN LAAR, H.H. 8015
VAN LEEUWEN, P.H. see LEEUWEN,
P.H., van
VAN MEETEREN, U. see MEETEREN,
U., van
VAN OORSCHOT, J.L.P. see OORSCHOT,
J.L.P., van
VAN SOEST, P.J. 7508
VAN STEVENINCK, R.F.M. 7252
VAN VOLKENBURGH, E. 8016
VAN WIJK, A.L.M. 8017
VARRIANO-MARSTON, E. 7200
VARTANIAN, N. 6621
VARTHA, E.W. 8018
VASAN, B.S. 7674
VASIĆ, G. 7239
VASILIU, M. 8019
VAVRINA, C.A. 7475
VEERANJANEYULU, K. 6908
VELEMÍNSKÝ, J. 7062
VENENI, M. 8020
VENKATARAYAPPA, T. 8021
VENKATESAN, V. 7674
VERETENNIKOV, A.V. 8022
VERMA, S.B. 7497
VESELOVA, T.V. 8023
VESELOVSKII, V.A. 8023
VIANE, R. 8024
VICENTE-CHANDLER, J. 6608
VICHERKOVÁ, M. 7307
VIDRAȘCU, P. 7215
VIEIRA da SILVA, J. 7631
VIENKEN, J. 7784
VIG, A.C. 7845
VIJAYA KUMAR, C.S.K. 7929, 8025,
8026
VINES, H.M. 7428
VINOPAL, R.T. 8027
VINTILĂ, R. 7271
VIRGIN, H.I. 7933
VIRMANI, S.M. 7857
VIRZO de SANTO, A. 8028
VITKOV, M. 6933
VIVOLI, J. 7425
VOLKENBURGH, E., van see VAN
VOLKENBURGH, E.
VOLKOVA, R.I. 8029, 8030
VOLODARSKII, N.I. 8031 see also
VOLODARSKY, N.I. 8032
VOROBEIKOV, G.A. 8033

PLANT INDEX

This index contains plant genera and types interesting as experimental material
for physiological, ecological and agricultural studies. The Latin plant names are
the main items which present the reference number. English names of the most common
plants are cross-indexed.

Alopecurus 6704, 7302, 7571

Altingia 8040

Alysicarpus 6677

Alyssum 7212, 7213, 7214

Amaranthus 6677, 6717, 6823, 6980,
7070, 7319, 7389, 7559, 7571, 7575,
7788, 7812

Ambrosia 6823, 7070, 7607, 7939,
8038

Amomum 8040

Amygdalus 7305

Ananas 7305, 7526, 7748

Anchusa 7212, 7214

Andropogon 7226, 7227, 7256

Anemone 6914

Anthoxanthum 7534

Antirrhinum 8055

Apium 6634, 6724, 6863, 6906,
7152, 7273

Apocynum 8092

apple see *Malus*

apricot see *Armeniaca*

Aquilegia 6914, 7344, 7345

Arabidopsis 6717, 6863, 7736

Arachis 7305, 7359, 7409, 7587,
7588, 7713, 7929

Arbutus 7954

Arctium 7465

Arctostaphylos 7536, 7960

Areca 7060

Arenaria 6809

Arenga 8128

Arisaema 6614

Aristida 6677

Armeniaca 7129, 7525, 7934, 7956.
7957, 8094, 8117

Artemisia 6717, 7010, 7070, 7212,
7213, 7214, 7519, 7524, 7793, 7794,
8065

Arum 6827

Asarum 6614

Asclepias 8092

ash see *Fraxinus* or *Sorbus*

Asparagus 7212, 7624

aspen see *Populus*

Aspidosperma 7879

Aster 7982

Astragalus 7212, 7213, 7214, 7612

Atriplex 6717, 6731, 7010, 7076,
7089, 7212, 7214, 7222, 7223, 7305,
7343, 7383, 7486, 7544, 7576, 7577,
7671, 7673, 7704, 7710, 7735, 7802,
8007, 8078, 8091

Avena 6622, 6751, 6920, 6948,
6994, 7023, 7055, 7070, 7081, 7083,
7140, 7203, 7212, 7225, 7343, 7435,
7440, 7796, 7910, 7930, 7968

Avicennia 6658

B

Bacteria
    Escherichia 6654, 7769, 8027
    Halobacterium 7355, 7387
    Pseudomonas 6971, 8099
    Rhizobium 7400
    Rhodopseudomonas 7547
    Rhodospirillum 7861, 7862
    Salmonella 6897, 7298

banana see *Musa*

Banksia 6863

barley see *Hordeum*

Batis 6647

Bauhinia 7692

bean see *Phaseolus*

beech see *Fagus*

bermudagrass see *Cynodon*

Beta 6640, 6802, 6804, 6829, 6830, 6832, 6863, 6921, 6948, 6969, 7050, 7070, 7207, 7293, 7299, 7305, 7323, 7343, 7370, 7373, 7435, 7473, 7544, 7571, 7645, 7761, 7792, 7801, 7853, 7920, 7966, 7995, 7996, 8005, 8081

Betula 6673, 6777, 6836, 6837, 7002, 7019, 7305, 7343, 7555, 7578, 7579, 7621, 7791, 7995, 7996, 8012

birch see *Betula*

Borassus 7060

Borya 6905, 7035

Bouteloua 7070, 7083, 7269, 8082

Brachiaria 7226, 7227

Brachypodium 7733

Brassica 6610, 6643, 6723, 6815, 6855, 6863, 6904, 7035, 7070, 7142, 7273, 7299, 7305, 7313, 7419, 7627, 7702, 7703, 7817, 7822, 7870, 7915, 7916, 7925, 8020

broadleaf see *Hyptis*

brome grass see *Bromus*

Bromus 6650, 7070, 7302, 7508, 7733

Bryophyllum 7343

Bryophyta
    Acrocladium 7100
    Atrichum 8114
    Barbula 7738
    Bryum 6635
    Diphyscium 8114
    Funaria 7531
    Grimmia 6635, 7738
    Lophocolea 8114
    Mnium 7544
    Nowellia 8114
    Polytrichum 6873
    Sphagnum 7079
    Tortula 6635, 7907
    Trichocolea 8114
    Weissia 6635

Bumelia 7911

C

cacao see *Theobroma*

Caesalpinia 7692, 7911

Cajanus 6688

Cakile 7212, 7214

Calamagrostis 6997, 7302, 7649, 7650

Callianthemum 6914

Callistephus 8055

Calluna 7079

Caltha 6914, 7343

Camellia 7557, 7561

Camissonia 7486, 7802

Canavalia 7333

Canna 7343

Cannabis 6960, 7868

Capparis 7386

Capsicum 6988, 7163, 7305, 7328, 7803, 8126

Carex 6697, 6838, 6863, 7068, 7212, 7214, 7302, 7304, 7305, 7343, 7371, 7649, 7650, 8065

Carissa 7397, 7683, 7684

Carpinus 6974, 7020, 7184, 7185, 7621, 7998

carrot see *Daucus*

Carthamus 7645, 7849

Carya 6614, 6754

Caryota 8128

Casearia 7911

Cassia 7692

Castanea 6923, 7305

Castanopsis 8040

Ceanothus 7536, 7556, 7775

cedar see *Tamarix*

Cedrela 7192

Celosia 6677, 7000, 8055

Celtis 6614

Cenchrus 6650, 6677, 7383, 7995, 7996, 8073

Centaurea 7214, 7887

Cerastium 6704

Cerasus 6613

Ceratonia 7891

Cereus 7453

Chamaecyparis 7530, 8127

Chenopodium 7070, 7272, 7273, 7946, 8126

cherry see *Cerasus*

Chloris 6650, 6677

Chlorophytum 8116

Chrysanthemum 6632, 7212, 7213, 7214, 7291, 7519, 7574

Cicer 6688, 7410, 7512, 7523

Cichorium 7212, 7214

Cissus 8028

Citrullus 7070, 7343

Citrus 6717, 6744, 6755, 6835, 6966, 7006, 7025, 7028, 7032, 7093, 7305, 7306, 7365, 7415, 7467, 7481, 7482, 7491, 7492, 7505, 7561, 7645, 7801, 7918, 8108

Clematis 6914

Cleome 6677

clover see *Trifolium*

Clusia 7879

Cocos 7035, 7060, 7343

Coffea 7566

Coix 7300, 7942

Coleus 8055

Commelina 6912, 6942, 6943, 6946, 7210, 7346, 7402, 7403, 7466, 7666, 7812, 7974, 8038, 8060, 8071

Consolida 6914

Convolvulus 6614, 6677, 7212, 7213, 7214

Coptis 6914

Corchorus 6677, 6688, 7405, 7605, 7685, 7686, 7804

Cordia 6756, 7660, 7911

Coreopsis 7645

cornel see *Cornus*

Cornus 6780, 6831, 7130, 7156, 7183, 7305, 7599, 7712, 8037

Corylus 6998

cotton see *Gossypium*

Cotula 7519

cowpea see *Vigna*

Cranbe 7212, 7214

cranberry see *Vaccinium*

Crataegus 7156, 7183

Crotalaria 6677, 7645

Croton 7911

Cryptomeria 7530

cucumber see *Cucumis*

Cucumis 6624, 6677, 6789, 6956, 6957, 7035, 7070, 7168, 7257, 7305, 7553, 7559, 7580, 7691, 7836, 7878, 8053, 8099

Cucurbita 6863, 7133, 7528, 7801, 7802

Cyclamen 7801

Cynodon 6650, 7343

Ficus 6756, 7305, 7625, 7626

Filipendula 6701, 6702

fir see *Abies*

Flacourtia 7397, 7683, 7684, 8040

Fragaria 6628, 6842, 7277, 7544

Fraxinus 6614, 6754, 6923, 7020, 7066, 7305, 7469, 7555, 7621

French bean see *Phaseolus*

Fuchsia 6827

Fungi
    Alternaria 8025, 8026
    Botrytis 7124, 7125
    Cercospora 7391
    Cytospora 7934
    Erysiphe 6662, 7651, 7652, 7653
    Fusarium 6718, 6973, 7172, 7244
    Heterobasidion 7322
    Leveillula 7328
    Macrophomina 7244
    Penicillium 7160
    Peronospora 7190, 7191
    Physoderma 7339
    Phytophthora 7064, 7244, 7288, 7396, 8120
    Pisolithus 6931
    Podosphaera 7427
    Puccinia 7899, 7929, 8013
    Pythium 7244, 7369
    Rhizoctonia 7244
    Saccharomyces 6782
    Sclerotium 6894
    Septoria 8013
    Streptomyces 7007
    Verticillium 6718, 6884

G

Galenia 8068

Gardenia 7095

Garrya 7960

Gaultheria 8040

Gerbera 7444, 7445

Geum 7035, 7304

Ginkgo 7343, 7549, 7621

Gisekia 6677

Glaucium 7454, 7455

Glechoma 6646

Gleditsia 7945

Glyceria 7343

Glycine 6631, 6644, 6696, 6720, 6722, 6738, 6769, 6770, 6778, 6799, 6818, 6823, 6834, 6863, 6874, 6945, 6947, 7003, 7013, 7070, 7083, 7136, 7137, 7138, 7154, 7181, 7190, 7191, 7229, 7230, 7241, 7266, 7273, 7288, 7294, 7305, 7311, 7343, 7383, 7585, 7400, 7418, 7419, 7423, 7424, 7441, 7463, 7480, 7504, 7515, 7544, 7575, 7576, 7577, 7587, 7610, 7616, 7644, 7645, 7646, 7696, 7713, 7806, 7807, 7808, 7841, 7843, 7847, 7853, 7855, 7865, 7896, 7918, 7927, 7928, 7947, 7948, 7949, 7955, 7961, 7995, 7996, 8051, 8052, 8064, 8083

Gossypium 6611, 6696, 6765, 6786, 6863, 6875, 6876, 6907, 6909, 6945, 7024, 7070, 7134, 7135, 7138, 7229, 7248, 7305, 7374, 7375, 7383, 7411, 7442, 7459, 7544, 7576, 7622, 7631, 7637, 7645, 7663, 7751, 7766, 7801, 7850, 7881, 7918, 7947, 7971, 7997

Goupia 7660

grape vine see *Vitis*

groundnut see *Arachis*

H

Haematoxylon 7911

Halophila 6694

Haloxylon 7001

Hammada 6717, 7383, 7576, 7577, 7793, 7794, 7995, 7996

Haplopappus 8038

Hedera 6691, 6717, 6966, 7305, 7343

Hedysarum 7186

Larrea 6717, 6731, 7383, 7486, 7577, 7607, 7612, 7802

Lemna 6863, 6878

Lens 6688, 7244

Lepidium 7035

Lespedeza 7945

lettuce see *Lactuca*

Leucanthemum 7519

Lichenes
    Cladonia 7835, 7953
    Cladophora 6752
    Collema 7835
    Hypogymnia 7630
    Peltigera 6892, 7869
    Ramalina 6892, 7246, 7341
    Umbilicaria 7347

Ligustrum 7048, 7183, 7305, 7343

Linum 6960, 7144, 7449, 7719, 7853

Liquidambar 7813

Liriodendron 6753, 7305, 7693, 7712, 8037

loblolly pine see *Pinus*

Lolium 6650, 6656, 6766, 6779, 6828, 6951, 6960, 6995, 7011, 7231, 7232, 7240, 7265, 7305, 7356, 7480, 7534, 7544, 7563, 7568, 7571, 7593, 7654, 7743, 7799, 7870, 7920, 7921, 7985, 8018, 8041, 8046

Lonicera 6809

Lotus 7439, 7440, 7870

lupine see *Lupinus*

Lupinus 6643, 6696, 7583, 7603, 7604, 7633, 7758, 7870, 7913, 8038

Luzula 6704

Lychnis 7035

Lycium 7607

Lycopersicon 6687, 6774, 6775, 6776, 6789, 6863, 6880, 6881, 6884, 6949, 6960, 6988, 6993, 7000, 7051, 7096, 7152, 7153, 7201, 7262, 7275,

Lycopersicon (continued) 7305, 7343, 7429, 7540, 7559, 7569, 7638, 7645, 7679, 7801, 7881, 7893, 7918, 7943, 7958, 7959, 8002, 8056, 8100

M

Maba 7397, 7683, 7684

Magnolia 7130

maize see *Zea*

Malus 6680, 6910, 6996, 7159, 7171, 7305, 7325, 7329, 7331, 7343, 7422, 7427, 7467, 7485, 7618, 7746, 7995, 7996, 8110

Manihot 6863, 6871

maple see *Acer*

Marrubium 7213, 7214

Matricaria 7519

Medicago 6627, 6648, 6728, 6748, 7070, 7090, 7109, 7212, 7214, 7343, 7356, 7456, 7472, 7480, 7563, 7593, 7594, 7640, 7645, 7657, 7665, 7667, 7672, 7700, 7715, 7729, 7730, 7761, 7765, 7801, 7855, 7865, 7917, 7922, 8085, 8111

Melothria 6677

Mentha 6859

Mesembryanthemum 7577, 7671, 7982, 8080, 8126

Metrosideros 8075

millet see *Panicum*

Mimosa 7933

Miscanthus 6624

Mollugo 7270

Momordica 7967

Morus 7951

mosses see *Bryophyta*

mulberry see *Morus*

mung bean see *Vigna*

Pharbitis 7769

Phaseolus 6618, 6619, 6627, 6667,
6672, 6677, 6685, 6688, 6727, 6733,
6764, 6785, 6786, 6825, 6832, 6833,
6863, 6971, 7033, 7038, 7070, 7082,
7094, 7097, 7098, 7113, 7166, 7174,
7252, 7305, 7321, 7333, 7336, 7343,
7429, 7432, 7438, 7447, 7484, 7523,
7567, 7570, 7571, 7576, 7577, 7587,
7598, 7602, 7607, 7628, 7637, 7645,
7673, 7760, 7809, 7810, 7811, 7824,
7825, 7833, 7850, 7853, 7854, 7863,
7882, 7883, 7884, 7897, 7937, 7987,
8016, 8038, 8045, 8057

Philodendron 7624

Phleum 6828, 7562, 7870

Phragmites 7677, 7729, 7747

Phyllanthus 7911

Picea 6811, 6891, 6893, 7019,
7092, 7093, 7126, 7131, 7173, 7187,
7209, 7259, 7267, 7305, 7343, 7362,
7383, 7621, 7998, 8022

pine see *Pinus*

Pinus 6660, 6664, 6665, 6693,
6697, 6712, 6730, 6753, 6777, 6794,
6845, 6863, 6885, 6893, 6965, 6999,
7035, 7077, 7123, 7145, 7147, 7187,
7267, 7305, 7322, 7343, 7377, 7378,
7450, 7478, 7517, 7518, 7530, 7572,
7621, 7723, 7739, 7740, 7741, 7742,
7821, 7848, 7876, 7951, 7998, 8012,
8035, 8043, 8061

Piper 7529, 7958, 7959

Pirus 6843, 6863, 7171, 7261,
7305, 8098

Pistacia 6926

Pisum 6627, 6662, 6679, 6703,
6721, 6757, 6854, 6863, 6935, 6938,
6939, 6940, 6960, 7128, 7178, 7188,
7199, 7212, 7250, 7343, 7358, 7438,
7480, 7507, 7544, 7636, 7645, 7675,
7676, 7687, 7745, 7767, 7768, 7844,
7912, 7914, 7940, 7995, 7996, 8097

Plantago 6863, 6888, 6889, 6985,
6986, 7216, 7323, 7534

Platanus 6614, 6823, 7254

Plectranthus 7544

Poa 6650, 6697, 7068, 7070, 7302,
7544, 8067

Polygonum 6614, 6806, 7070

Populus 6614, 6781, 6790, 6823,
6839, 6841, 6928, 6930, 7020, 7073,
7186, 7268, 7305, 7343, 7447, 7460,
7544, 7555, 7585, 7586, 7621, 7645,
7829, 7858, 7859

Portulaca 7035, 7070, 7237, 7296,
7297, 7571

Potamogeton 7343

potato see *Solanum*

Primula 6838

Prosopis 6740, 7047, 7120, 7430,
7488, 7621

Prunus 6614, 6753, 7021, 7171,
7305, 7343, 7383, 7620, 7660, 7793,
7794, 7999, 8009, 8038

Pseudotsuga 6661, 6736, 7093,
7305, 7350, 7352, 7378, 7535, 7621,
7739

Pteridophyta
    Adianthum 7380
    Aneimia 7380
    Aspidium 7380
    Asplenium 6844, 7380, 7641,
    8024
    Blechnum 7380
    Dipteris 7274
    Drynaria 7380
    Dryopteris 7380
    Elaphoglossum 7380
    Equisetum 6802
    Humata 8040
    Lycopodium 6802
    Onoclea 6879
    Phyllitis 7380, 8024
    Platycerium 7380
    Polypodium 7035, 7380
    Polystichum 7380
    Pteridium 7711
    Pteris 7380
    Selaginella 6802, 6905, 6919,
    6970, 8040, 8065

Puccinellia 6704, 7903, 7921,
8050, 8091

Pulicaria 6677

Pulmonaria 6975

Triticum (continued) 7720, 7734,
7750, 7759, 7770, 7771, 7772, 7798,
7801, 7816, 7817, 7821, 7839, 7844,
7850, 7851, 7852, 7853, 7865, 7867,
7881, 7886, 7892, 7899, 7900, 7901,
7910, 7914, 7918, 7920, 7925, 7926,
7930, 7933, 7944, 7964, 7969, 7972,
7977, 7983, 7984, 7995, 7996, 8003,
8013, 8019, 8025, 8026, 8033, 8036,
8089

Trollius 6914

Tropaeolum 7466

Tsuga 6707, 7045, 7245, 7305,
8127

Tulipa 8038, 8116

U

Ulmus 6614, 6646, 7305, 7549,
7555, 7645

Urtica 6614

V

Vaccinium 6911, 7306, 7890

Verbascum 7212

Veronica 6677

Viburnum 7130, 7156, 7716,
7717, 8040

Vicia 6719, 6733, 6827, 6927,
6946, 7124, 7125, 7212, 7213, 7214,
7247, 7249, 7250, 7300, 7307, 7318,
7335, 7343, 7383, 7389, 7558, 7559,
7560, 7636, 7645, 7721, 7778, 7779,
7780, 7781, 7782, 7783, 7784, 7812,
7831, 8038, 8041, 8097, 8115, 8116,
8125

Vigna 6618, 6818, 7036, 7037,
7070, 7099, 7104, 7105, 7305, 7566,
7609, 7673, 7787, 7832, 7990, 7991,
7992, 7993

Vitis 6670, 6671, 6959, 7029,
7305, 7383, 7446, 7452, 7576, 7597,
7797, 7801, 7827, 7996, 7997

W

walnut see *Juglans*

Washingtonia 6917, 8128

wheat see *Triticum*

willow see *Salix*

X

Xanthium 7070, 7637, 7812, 8038,
8113

Xerophyta 7107

Y

Yucca 7536, 8014, 8041

Z

Zantedeschia 8116

Zea 6609, 6619, 6620, 6624, 6627,
6630, 6650, 6656, 6675, 6696, 6721,
6761, 6762, 6767, 6783, 6798, 6807,
6808, 6820, 6821, 6823, 6832, 6851,
6856, 6857, 6863, 6869, 6872, 6912,
6916, 6920, 6921, 6933, 6945, 6954,
6960, 6963, 6967, 6973, 6991, 7030,
7035, 7044, 7053, 7054, 7056, 7063,
7070, 7074, 7083, 7094, 7106, 7182,
7188, 7229, 7238, 7239, 7256, 7271,
7300, 7305, 7311, 7316, 7320, 7332,
7339, 7342, 7343, 7370, 7381, 7383,
7390, 7406, 7425, 7462, 7470, 7471,
7476, 7480, 7490, 7493, 7537, 7554,
7559, 7561, 7571, 7577, 7583, 7617,
7632, 7639, 7645, 7673, 7713, 7745,
7755, 7756, 7761, 7762, 7770, 7811,
7812, 7820, 7821, 7831, 7837, 7846,
7850, 7853, 7870, 7884, 7898, 7901,
7908, 7918, 7932, 7933, 7942, 7943,
7951, 7965, 7974, 7986, 7995, 7996,
7997, 8010, 8011, 8015, 8020, 8038,
8054, 8091, 8111, 8116, 8121, 8122

Zygophyllum 7383, 7576, 7793,
7794

SUBJECT INDEX

This index contains a selection of primary items chosen according to their interest
for water relation researchers and to their relative importance and occurrence.

A

Abaxial and adaxial epidermes 6611, 6618, 6650, 6662, 6684, 6737, 6739, 6793, 6794,
        6809, 6843, 6844, 6859, 6914, 6919, 6947, 7031, 7151, 7173, 7189, 7227, 7256,
        7263, 7276, 7305, 7333, 7405, 7419, 7461, 7520, 7521, 7522, 7557, 7582, 7585,
        7587, 7602, 7645, 7660, 7683, 7690, 7692, 7736, 7807, 7813, 7843, 7929, 7974,
        8066, 8067, 8092, 8128

Abscisic acid see Antitranspirants; Growth substances, hormones, inhibitors etc. ...

Absorption of water see Water absorption ...

Age of leaf see Age of plant ...; Leaf insertion level ...

Age of plant and conductance for water vapour and carbon dioxide transfer 6612, 6833,
        7241, 7421, 7485, 7881, 7883

Age of plant and transpiration 6619, 6621, 6642, 6643, 6700, 6737, 7049, 7603, 7657,
        7665, 7689, 7744, 7979, 7992

Age of plant and water absorption by plant 6700, 6773, 7886, 7897, 7905, 7910, 7996

Age of plant and water status in plant 6612, 6663, 6681, 6737, 6954, 7098, 7231,
        7234, 7241, 7336, 7495, 7603, 7735, 7775, 7852, 7868, 7910, 7990, 8106

Age of plant and water transport in plant 7309

Age of plant and wilting 6616, 6829, 6851, 6853, 7014, 7238, 7358, 7390, 7408, 7453,
        7496, 7509, 7621, 7886, 7905, 7910, 7919, 7970, 7996, 8101

Agrotechnics see Farming practices ...

Air-conditioning see Transpiration rate, methods, gasometric systems

Air-flow rate see Wind ...

Altitude, pressure and stomata 7305, 7524

Amino acids see Proteins, amino acids, nucleic acids ...

Amphistomatous leaves 7829

Anatomical structure and transpiration rate 6684

Anatomical structure and wilting 7453, 7996, 8062

Anatomical structure of epidermis 7519, 7586

Antibiotics and wilting 6970

Antitranspirants (see also Growth substances, hormones, inhibitors etc. ...) 6611,
        6707, 6871, 6888, 6907, 7252, 7291, 7368, 7467, 7476, 7574, 7609, 7734, 7859,
        7918, 8060

Assimilation see Carbon dioxide influx

Availability of soil water 6606, 6610, 6614, 6645, 6653, 6659, 6661, 6687, 6700,
        6701, 6707, 6711, 6712, 6726, 6735, 6745, 6766, 6767, 6769, 6773, 6777, 6790,
        6819, 6829, 6837, 6845, 6851, 6860, 6894, 6913, 6948, 6950, 6962, 6969, 6997,
        7012, 7041, 7061, 7083, 7109, 7111, 7117, 7126, 7138, 7141, 7142, 7165, 7197,
        7202, 7205, 7217, 7222, 7230, 7231, 7232, 7236, 7242, 7243, 7247, 7277, 7316,
        7317, 7356, 7396, 7410, 7423, 7424, 7433, 7434, 7435, 7442, 7456, 7457, 7459,
        7493, 7512, 7539, 7566, 7581, 7594, 7600, 7664, 7696, 7700, 7713, 7743, 7758,
        7759, 7766, 7805, 7815, 7820, 7845, 7857, 7886, 7897, 7910, 7915, 7916, 7918,
        7923, 7925, 7930, 7948, 7963, 7971, 7985, 7990, 7992, 7996, 7998, 8039, 8046,
        8065, 8069, 8093, 8111

B

Beta gauge see Water saturation deficit, methods

Biological clock see Diurnal changes

Books on plant water relation see General aspects ...

Bound water 6671, 6759, 7006, 7054, 7186, 7215, 7320, 7331, 7332, 7336, 7446, 7465,
        7525, 7528, 7538, 7677, 7712, 8073, 8086, 8090, 8096, 8098

Bound water, methods 7511

Boundary layer of leaf see Conductance for water vapour and carbon dioxide transfer,
        boundary layer of leaf

C

C_3, C_4, CAM pathways see Comparison of plants with different types of carbon me-
        tabolism

Canopy architecture see Irrigation ...; Precipitation, dew ...; Salinity ...;
        Water status in plant and canopy architecture

Canopy model see Model of canopy

Canopy water vapour profiles see Humidity of air, gradients in canopy

Carbohydrates and conductance for water vapour and carbon dioxide transfer 7808

Carbohydrates and stomata 7968

Carbohydrates and transpiration 7684

Carbohydrates and water status in plant 6682, 6787, 6926, 7233, 7871, 7872, 7873,
        7996, 8018, 8073

Carbohydrates and water transport in cells 7314

Carbohydrates and wilting 6682, 6829, 7213, 7233, 7452, 7823, 7965, 7996, 8018, 8073

Carbon dioxide and conductance for water vapour and carbon dioxide transfer 6889,
        7151, 7153, 7209, 7490, 7545, 7596, 7808, 7883, 7996

Carbon dioxide and stomata 6624, 6888, 6889, 6998, 7209, 7210, 7335, 7360, 7381, 7545, 7619, 7779, 7782, 7811, 7996

Carbon dioxide and transpiration 6623, 6624, 6823, 7092, 7151, 7349, 7360, 7374, 7381, 7389, 7561, 7860

Carbon dioxide and water status in plant 7360, 7851

Carbon dioxide and water transport in cells 7937

Carbon dioxide influx see Deuterium oxide, tritium oxide ...; Drought ...; Flooding ...; Humidity of air ...; Irrigation ...; Osmotically active substances ...; Precipitation, dew ...; Salinity ...; Soil moisture ...; Water status in plant and carbon dioxide influx

Carbon fixation pathways see Deuterium oxide, tritium oxide ...; Drought ...; Irrigation ...; Osmotically active substances ...; Salinity ...; Soil moisture ...; Water status in plant and carbon fixation pathways

Carbowax see Osmotically active substances ...

Carotenoids see Drought ...; Osmotically active substances ...; Salinity ...; Soil moisture ...; Water status in plant and carotenoids

Chlorophyll see Deuterium oxide, tritium oxide ...; Drought ...; Flooding ...; Humidity of air ...; Osmotically active substances ...; Precipitation, dew ...; Salinity ...; Soil moisture ...; Water status in plant and chlorophyll

Chloroplasts see Osmotically active substances ...; Salinity ...; Soil moisture ...; Water status in plant and chloroplasts

Clock, biological see Diurnal changes ...

CO$_2$ see Carbon dioxide ...

Comparison of plants with different types of carbon metabolism see Conductance for water vapour and carbon dioxide transfer ...; Stomata ...; Transpiration rate ...; Water status in plant, comparison of plants with different types of carbon metabolism

Conductance for carbon dioxide transfer, epidermis 6620, 6658, 6685, 6771, 6832, 6839, 6841, 6877, 6908, 7270, 7305, 7341, 7343, 7421, 7612, 7654, 7764, 7807, 7808, 7876, 7894, 8076, 8104

Conductance for carbon dioxide transfer, mesophyll (intracellular) 6620, 6685, 6698, 6833, 6839, 6840, 6841, 6876, 6915, 6917, 6947, 7008, 7113, 7153, 7170, 7269, 7270, 7305, 7319, 7362, 7370, 7374, 7375, 7384, 7420, 7421, 7544, 7545, 7578, 7612, 7616, 7654, 7689, 7754, 7807, 7955, 7957, 7961, 7976, 7996, 8037

Conductance for water vapour and carbon dioxide transfer see Age of plant ...; Carbohydrates ...; Carbon dioxide ...; Conductance for carbon dioxide transfer ...; Conductance for water vapour transfer ...; Cultivars ...; Deuterium oxide, tritium oxide ...; Drought ...; Ecotypes ...; Enzymes ...; Farming practices ...; Flooding ...; Genetics ...; Growth substances, hormones, inhibitors etc. ...; Humidity of air ...; Irradiance ...; Irrigation ...; Leaf insertion level ...; Mineral elements ...; Osmotically active substances ...; Oxygen ...; Pathogens ...; Pesticides, herbicides ...; Pollutants, ozone ...; Precipitation, dew ...; Salinity ...; Soil moisture ...; Taxons ...; Temperature ...; Water status in plant ...; Wind and conductance for water vapour and carbon dioxide transfer

Conductance for water vapour and carbon dioxide transfer, above canopy 7078

Diurnal changes see Conductance for water vapour and carbon dioxide transfer, ...;
      Stomatal aperture, ...; Transpiration rate, ...; Water absorption by plant,
      ...; Water status in plant, ...; Water transport in plant, ...; Wilting,
      diurnal changes

D_2O, T_2O see Deuterium oxide, tritium oxide ...

Drainage see Farming practices ...

Drought and carbon dioxide influx 6786, 7094, 7231, 7292, 7296, 7297, 7312, 7343,
      7351, 7357, 7386, 7421, 7486, 7723, 7755, 7794, 7802, 7918, 7996

Drought and carbon fixation pathways 7292, 7296

Drought and carotenoids 6747

Drought and chlorophyll 6747, 6762, 7703, 7802

Drought and conductance for water vapour and carbon dioxide transfer 6611, 7002,
      7130, 7231, 7232, 7268, 7351, 7377, 7449, 7582, 7621, 7645, 7723, 7996

Drought and electron transport chain 8032

Drought and growth and productivity 6625, 6636, 6651, 6659, 6682, 6696, 6783, 6792,
      6822, 6834, 6853, 6869, 6973, 6992, 7014, 7015, 7016, 7041, 7051, 7084, 7116,
      7117, 7135, 7149, 7231, 7235, 7238, 7268, 7279, 7292, 7312, 7326, 7327, 7334,
      7356, 7357, 7408, 7449, 7452, 7483, 7503, 7509, 7621, 7663, 7668, 7691, 7696,
      7702, 7709, 7730, 7799, 7801, 7913, 7918, 7919, 7938, 7970, 7971, 7990, 7991,
      7993, 7996, 8059

Drought and leaf anatomy 6682, 6696, 6853, 6901, 7231, 7232, 7268, 7344, 7360, 7452,
      7737, 7996

Drought and respiration 6786, 7117, 7231, 7357

Drought and stomata 6611, 6696, 6792, 7232, 7360, 7377, 7449, 7645, 7918, 7996

Drought and transpiration 6619, 7014, 7105, 7231, 7268, 7296, 7334, 7351, 7360, 7377,
      7535, 7657, 7723, 7755, 7865, 7918, 7996

Drought and water absorption by plant 7142, 7232

Drought and water status in plant 6611, 6648, 6681, 6926, 7002, 7186, 7232, 7351,
      7358, 7360, 7503, 7582, 7621, 7755, 7895, 7905, 7996, 8073

Drought and water transport in cells 7531, 7769

Drought and wilting 6730, 6935, 6952, 7228, 7992, 7996

Drought resistance 6651, 6660, 6682, 6725, 6730, 6743, 6761, 6768, 6792, 6808, 6821,
      6836, 6899, 6900, 6901, 6904, 6905, 6910, 6958, 6993, 7017, 7035, 7041, 7052,
      7083, 7102, 7105, 7106, 7116, 7117, 7118, 7149, 7183, 7186, 7212, 7213, 7214,
      7215, 7219, 7235, 7236, 7245, 7279, 7312, 7338, 7358, 7364, 7390, 7399, 7439,
      7495, 7535, 7542, 7594, 7621, 7631, 7661, 7663, 7701, 7702, 7704, 7710, 7712,
      7755, 7770, 7789, 7790, 7797, 7822, 7904, 7905, 7907, 7918, 7923, 7945, 7965,
      7988, 7990, 7991, 7992, 7993, 7996, 8002, 8003, 8038, 8062, 8068, 8119

Drought resistance, genetics see Genetics and drought resistance

Drought resistance, methods 7583

Dry matter production see Growth and productivity ...

E

Ear removal see Defoliation, decapitation, ear, root removal ...

Ecotypes, geographical types and conductance for water vapour and carbon dioxide
    transfer 6839, 6889, 7305, 7383, 7996

Ecotypes, geographical types and stomata 6889, 7383, 7524, 7996

Ecotypes, geographical types and transpiration 6839, 7088, 7302, 7870, 7979, 8001

Ecotypes, geographical types and water status in plant 7015, 7214

Ecotypes, geographical types and wilting 6660, 6696, 7015, 7212, 7302, 7535

Electron transport chain see Drought ...; Humidity of air ...; Irrigation ...;
        Osmotically active substances ...; Salinity ...; Soil moisture ...; Water
        status in plant and electron transport chain

Enzyme inhibitors and stomata 7666

Enzyme inhibitors and water absorption by plant 6798

Enzymes and water status in plant 7260

Enzymes and wilting 6884, 7623, 7982

Epidermal conductance (resistance) see Conductance for water vapour and carbon
        dioxide transfer, epidermis

Epidermis see Abaxial and adaxial epidermes; Anatomical structure of epidermis;
        Conductance for carbon dioxide transfer, epidermis; Conductance for water
        vapour transfer, epidermis; Stomata and epidermis, heterogeneity of single
        leaf blade

Evaporation 6687, 6700, 6811, 6867, 6920, 6956, 6963, 7021, 7029, 7042, 7080, 7089,
        7111, 7162, 7205, 7386, 7392, 7399, 7423, 7424, 7426, 7433, 7434, 7435, 7456,
        7470, 7493, 7525, 7613, 7614, 7684, 7743, 7747, 7763, 7764, 7938, 7943, 7969,
        7979, 7992

Evapotranspiration 6608, 6621, 6627, 6637, 6642, 6643, 6645, 6661, 6684, 6697, 6700,
        6711, 6713, 6748, 6811, 6829, 6837, 6936, 6956, 6962, 6963, 6969, 6997, 7011,
        7042, 7111, 7114, 7115, 7126, 7138, 7162, 7193, 7198, 7205, 7217, 7224, 7230,
        7283, 7311, 7316, 7318, 7342, 7392, 7410, 7416, 7424, 7455, 7457, 7470, 7471,
        7473, 7474, 7483, 7497, 7515, 7536, 7539, 7543, 7565, 7593, 7605, 7649, 7650,
        7696, 7720, 7759, 7763, 7771, 7772, 7799, 7800, 7801, 7804, 7805, 7832, 7865,
        7880, 7884, 7918, 7932, 7943, 7948, 7978, 7979, 7992, 8012, 8118

Evapotranspiration, methods, evaporimeters and lysimeters 6811, 6816, 6817, 6835,
        6951, 6953, 7138, 7224, 7828, 7943, 7978

Evapotranspiration, methods, other 7114, 7115, 7224, 7605, 7649, 7650, 7747, 7763,
        8043

Exhaust gases see Pollutants, ozone ...

Exposure chamber see Transpiration chambers

Extension see Growth and productivity ...

Exudation see Root pressure, exudation

F

Farming practices and conductance for water vapour and carbon dioxide transfer 6625, 6664, 6680, 7092, 7278

Farming practices and stomata 6793, 6891, 7092, 7562, 7563

Farming practices and transpiration 6664, 6891, 7092, 7179, 7195, 7416, 7934

Farming practices and water absorption by plant 6767, 6860, 6950, 7566, 7581, 8045, 8120

Farming practices and water status in plant 6625, 6767, 7092, 7278, 7674

Farming practices and wilting 6606, 7518

Fatty acids see Lipids, fatty acids ...

Flooding and carbon dioxide influx 7983, 8022

Flooding and chlorophyll 7669, 7984

Flooding and conductance for water vapour and carbon dioxide transfer 6774, 7033, 7377

Flooding and growth and productivity 6683, 6703, 6796, 6815, 6816, 6817, 6972, 6989, 7066, 7135, 7201, 7240, 7356, 7367, 7668, 7670, 7680, 7796, 7833, 7983, 8019, 8033, 8056, 8057, 8103

Flooding and leaf anatomy 6703, 7034, 7983, 8033

Flooding and stomata 7066, 7377, 7881

Flooding and transpiration 7066, 7377, 7755, 7983, 8056

Flooding and water absorption by plant 7066

Flooding and water status in plant 6774, 7755, 7983

Fluorine see Pollutants, ozone ...

Foliage see Canopy ...

Frost resistance 6634, 6730, 6736, 6778, 6797, 6810, 6842, 6849, 6923, 7023, 7048, 7101, 7158, 7267, 7401, 7406, 7446, 7590, 7594, 7597, 7599, 7623, 7717, 7718, 7859, 7904, 7922, 8090, 8094, 8106

Frost resistance, methods 7101

Fungal diseases see Pathogens ...

G

Gas exchange see Carbon dioxide influx ...; Transpiration ...;

Gasometric methods see Transpiration rate, methods, gasometric systems

General aspects on plant water relations 6638, 6726, 6729, 6838, 7026, 7071, 7111, 7164, 7217, 7292, 7343, 7360, 7361, 7424, 7697, 7722, 7801, 7864, 7918, 7996

Genetics and conductance for water vapour and carbon dioxide transfer 7027

Genetics and drought resistance 6725, 6743, 7035, 7117, 7221, 7236, 7390, 7770, 7907, 7918

Genetics and stomata 6720, 6800, 6806, 6855, 6959, 7027, 7173, 7189, 7206, 7338, 7404, 7422, 7489, 7586, 7648, 7692, 7736, 7931

Genetics and transpiration 6800, 7027, 7342, 7681

Genetics and water absorption by plant 7041, 7197, 7850

Genetics and water status in plant 6630, 7062, 7118, 7197, 7248, 7494, 7495, 7678, 7996

Growth and productivity see Drought ...; Flooding ...; Humidity of air ...; Irrigation ...; Osmotically active substances ...; Precipitation, dew ...; Salinity ...; Soil moisture ...; Water status in plant and growth and productivity

Growth substances, hormones, inhibitors etc. and conductance for water vapour and carbon dioxide transfer 6611, 6780, 6871, 6946, 7055, 7151, 7258, 7384, 7397, 7431, 7609, 7808, 7858, 7864, 8093

Growth substances, hormones, inhibitors etc. and stomata 6611, 6615, 6656, 6871, 6912, 6946, 6988, 7055, 7210, 7218, 7300, 7346, 7397, 7429, 7466, 7557, 7574, 7659, 7684, 7812, 7864, 7968, 7996, 8038, 8060, 8092

Growth substances, hormones, inhibitors etc. and transpiration 6808, 6907, 6988, 7151, 7240, 7295, 7300, 7431, 7557, 7561

Growth substances, hormones, inhibitors etc. and water status in plant 6611, 6656, 6808, 6909, 6912, 6946, 7257, 7384, 7450, 7552, 7637, 7859, 7864, 7996, 8038, 8093, 8096

Growth substances, hormones, inhibitors etc. and water transport in cells 7417, 7445, 7714, 7769

Growth substances, hormones, inhibitors etc. and water transport in plant 7417

Growth substances, hormones, inhibitors etc. and wilting 6611, 6615, 6634, 6656, 6665, 6820, 6822, 6909, 6912, 7017, 7035, 7175, 7201, 7384, 7467, 7535, 7552, 7574, 7625, 7661, 7858, 7859, 7864, 7923, 7996, 8038, 8089, 8113

Guard cells see Stomata ...; Stomatal ...;

Guttation 6870, 6890

H

H$_2$ isotopes see Deuterium oxide, tritium oxide ...;

Heterogeneity of single leaf blade see Conductance for water vapour and carbon dioxide transfer ...; Stomata and epidermis, ...; Transpiration ...; Water status in plant, heterogeneity of single leaf blade

Hormones see Growth substances, hormones, inhibitors etc. ...

Humidity of air and carbon dioxide influx 6620, 6624, 6726, 6925, 6998, 7065, 7104, 7113, 7254, 7579, 7809, 7841

Humidity of air and chlorophyll 6744, 7835, 8117

Humidity of air and conductance for water vapour and carbon dioxide transfer 6620,
       6726, 6733, 6889, 6934, 6974, 6998, 7092, 7104, 7209, 7365, 7383, 7431, 7501,
       7502, 7579, 7695, 7739, 7809, 7937, 7996, 8127

Humidity of air and electron transport chain 7835

Humidity of air and growth and productivity 6621, 6777, 7030, 7348, 7438, 7451, 7480,
       7635, 7795, 7970, 8114

Humidity of air and leaf anatomy 6621, 6745, 6936, 7022, 7360, 7544, 7996

Humidity of air and stomata 6889, 6974, 6998, 7092, 7209, 7335, 7343, 7360, 7383,
       7402, 7809, 7996, 8000, 8127

Humidity of air and transpiration 6621, 6649, 6726, 6733, 6956, 6998, 7021, 7047,
       7105, 7209, 7253, 7360, 7365, 7431, 7480, 7579, 7620, 7649, 7689, 7739, 7996

Humidity of air and water absorption by plant 7021

Humidity of air and water status in plant 6726, 7092, 7124, 7131, 7209, 7232, 7348,
       7360, 7365, 7488, 7695, 7735, 7739

Humidity of air and water transport in plant 8002

Humidity of air and wilting 6640, 7352, 7918

Humidity of air, gradients in canopy 6705, 6920, 7764, 7800, 7908

Humidity of air in leaf intercellular spaces see Transpiration rate, gradients of
       air humidity in leaf intercellular spaces

Hydration level see Water status in plant ...

Hydraulic conductivity see Water transport in plant ...

Hydroactive closure of stomata see Water status in plant and stomata

Hypostomatous leaves 6959

Hydrogen isotopes see Deuterium oxide, tritium oxide ...

I

Ideotype see Model ...

Infra-red gas analyser see Transpiration rate, methods, infra-red gas analysers

Inhibitors see Growth substances, hormones, inhibitors etc. ...

Insertion level see Leaf insertion level ...

Integrated transpiration see Transpiration, integrated

Intercellular spaces see Conductance for water vapour and carbon dioxide transfer,
       intercellular spaces; Transpiration rate, gradients of air humidity in leaf
       intercellular spaces

Irradiance and conductance for water vapour and carbon dioxide transfer 6620, 6647,
       6698, 6726, 6733, 6779, 6840, 6889, 6932, 6934, 6974, 6987, 7008, 7047, 7092,
       7151, 7209, 7225, 7343, 7351, 7358, 7362, 7419, 7449, 7490, 7492, 7540, 7544,
       7545, 7556, 7558, 7578, 7587, 7596, 7621, 7645, 7646, 7654, 7808, 7811, 7877,
       7883, 7939, 7961, 7996, 8010, 8037, 8078, 8127

L

Latent heat transfer see Transpiration rate, theoretical background

Leaf anatomy see Drought ...; Flooding ...; Humidity of air ...; Irrigation ...;
        Osmotically active substances ...; Precipitation, dew ...; Salinity ...;
        Soil moisture ...; Water status in plant and leaf anatomy

Leaf chambers see Transpiration chambers

Leaf conductance (resistance) see Conductance for water vapour and carbon dioxide
        transfer, stomata

Leaf insertion level and conductance for water vapour and carbon dioxide transfer
        6611, 6876, 6975, 6977, 7002, 7073, 7276, 7420, 7485, 7616, 7645, 7689, 7711,
        7882, 7955, 8083, 8128

Leaf insertion level and stomata 6611, 6756, 6793, 6855, 6959, 7173, 7276, 7405,
        7521, 7522, 7629, 7645, 7690, 7787, 7882, 7968, 7987, 8066, 8105·

Leaf insertion level and transpiration 6793, 6876, 7002, 7464, 8010, 8011, 8040,
        8051

Leaf insertion level and water status in plant 6611, 6632, 6769, 6836, 6909, 6975,
        7002, 7163, 7233, 7234, 7331, 7458, 7465, 7679, 7712, 7734, 7745, 7798, 7848,
        7851, 7898, 7900, 8098

Leaf insertion level and wilting 6682, 6696, 6909, 7233, 7234, 7352, 7898, 7982,
        7996

Leaf surface, waxes and trichomes 6669, 6739, 6793, 6794, 6809, 6843, 6968,·7015,
        7028, 7060, 7256, 7276, 7333, 7381, 7422, 7448, 7519, 7520, 7521, 7522, 7656,
        7660, 7748, 7814, 7899, 7996, 8092

Leaf surface, waxes and trichomes, seasonal changes 6974, 7028, 7259, 8127

Light see Irradiance ...

Lipids, fatty acids and water transport in cells 7323

Lysimeters see Evapotranspiration, methods, evaporimeters and lysimeters

M

Mannitol see Osmotically active substances ...

Mass flow see Water transport in plant ...

Mathematical model see Model; Model of canopy

Matric potential in plant tissue (see also Water status in plant ...) 7601, 7712,
        7821

Matric potential in substrate 6687, 6772, 6894, 6997, 7064, 7073, 7130, 7136, 7217,
        7247, 7358, 7369, 7396, 7432, 7598, 7600, 7632, 7696, 7758, 7916, 8120,

Matric potential, methods 7758, 7821

Mesophyll conductance see Conductance for carbon dioxide transfer, mesophyll
        (intracellular)

Mineral elements and conductance for water vapour and carbon dioxide transfer 6766,
        6889, 7013, 7033, 7281, 7375, 7549, 7550, 7638, 7797, 8104

Osmotically active substances and conductance for water vapour and carbon dioxide
    transfer 7013, 7163, 7225, 7626, 7691

Osmotically active substances and electron transport chain 6709, 6715, 6854, 6938,
    6940, 7121, 7188, 7366, 7718, 7745

Osmotically active substances and growth and productivity 6678, 6704, 6718, 6758,
    6770, 6782, 6862, 6881, 6895, 7013, 7038, 7172, 7272, 7273, 7348, 7514, 7672,
    7675, 7691, 7704, 7756, 7757, 7826, 7838, 7891, 7962, 8055, 8063

Osmotically active substances and leaf anatomy 8055

Osmotically active substances and productivity of algae 6795, 7889

Osmotically active substances and respiration 7286, 7631, 7823, 8063

Osmotically active substances and stomata   7225, 7782

Osmotically active substances and transpiration 7225, 7553

Osmotically active substances and water absorption by plant 6880, 7294, 7458, 7487,
    7554, 8122

Osmotically active substances and water status in plant 6770, 6782, 6784, 7005,
    7054, 7152, 7358, 7395, 7554, 7626, 7675, 7676, 7704, 7982, 8080

Osmotically active substances and water transport in cells 6708, 7395, 7714

Osmotically active substances and wilting 6878, 6980, 7013, 7129, 7358, 7569, 7570,
    7625, 7676, 7823, 7826, 7971, 7982

Osmotic potential in plant tissue (see also Water status in plant ...) 6609, 6626,
    6656, 6660, 6662, 6663, 6681, 6682, 6696, 6726, 6732, 6770, 6782, 6784, 6787,
    6896, 6899, 6900, 6910, 6926, 7006, 7087, 7092, 7093, 7110, 7132, 7156, 7158,
    7163, 7174, 7197, 7214, 7216, 7232, 7233, 7234, 7235, 7245, 7248, 7255, 7257,
    7260, 7278, 7285, 7289, 7290, 7301, 7312, 7343, 7358, 7369, 7378, 7418, 7419,
    7427, 7442, 7444, 7458, 7469, 7476, 7494, 7495, 7553, 7576, 7601, 7615, 7637,
    7671, 7675, 7676, 7677, 7704, 7712, 7735, 7737, 7755, 7851, 7852, 7871, 7872,
    7874, 7888, 7904, 7905, 7906, 7921, 7922, 7941, 7950, 7990, 7995, 7996, 8005,
    8014, 8027, 8054, 8055, 8073, 8088, 8126

Osmotic potential in substrate 6906, 7005, 7013, 7152, 7216, 7286, 7287, 7373, 7590,
    7394, 7536, 7554, 7598, 7691, 7758, 7769, 7786, 7898, 7923, 7971, 8070, 8120

Osmotic potential, methods 7758, 7950, 8054

Oxygen and conductance for water vapour and carbon dioxide transfer 7319, 7616, 7881

Oxygen and stomata 7275, 7360, 7881

Oxygen and transpiration 6623, 6624, 6887, 7166, 7360, 7374, 7389, 7559, 7561

Oxygen and water absorption by plant 6887, 7166, 7881

Oxygen and water status in plant 6941, 7360

Oxygen and wilting 6887

Ozone see Pollutants, ozone ...

P

Pathogens and conductance for water vapour and carbon dioxide transfer 6662, 7463

Pathogens and stomata 6662, 7328, 7803, 7899, 8026, 8099

Pathogens and transpiration 6662, 7095, 7187, 7288, 7468, 7651, 7652, 7653, 7929, 7934, 8013, 8025

Pathogens and water absorption by plant 6790, 6894, 7396, 7651, 7652, 7653, 8108, 8120

Pathogens and water status in plant 7124, 7125, 7369, 7427, 7518, 8035

Pathogens and wilting 6730, 6851, 6884, 7125, 7129, 7651,' 7652, 7653, 8035

Permanent wilting see Wilting ...

Permeability see Water transport in cells, permeability

Pesticides, herbicides and conductance for water vapour and carbon dioxide transfer 7397, 7684, 7809

Pesticides, herbicides and stomata 7397, 7684, 7809, 7946, 8092

Pesticides, herbicides and transpiration 6619, 7397, 7447, 7571, 7684, 7810, 8052

Pesticides, herbicides and water absorption by plant 6890, 6948

Pesticides, herbicides and water status in plant 6846, 7406

Pesticides, herbicides and water transport in plant 7447, 7810

Pesticides, herbicides and wilting 6619, 6846, 7447

pH and stomata 7300, 7346, 8071

Photorespiration see Salinity and photorespiration

Photosynthesis see Carbon dioxide influx ...

Photosynthesis - transpiration ratio see Productivity of transpiration

Phytopathology see Pathogens ...

Plasmolysis see Water transport in cells, plasmolysis

Pollutants, ozone and conductance for water vapour and carbon dioxide transfer 6685, 6733, 6978, 7058, 7504, 7549, 7550, 7827, 8075, 8077, 8078

Pollutants, ozone and stomata 6809, 6937, 6979, 7058, 7204, 7300, 7549, 7813, 7814, 8077

Pollutants, ozone and transpiration 6685, 6733, 7039, 7271, 7300, 7525, 8076

Pollutants, ozone and water status in plant 7271, 7525

Pollutants, ozone and water transport in cells 7769

Pollutants, ozone and wilting 7525, 7549

Polyethylene glycol see Osmotically active substances ...

Porometer see Stomatal aperture, methods, diffusion porometers; Stomatal aperture,
      methods, viscous flow porometers

Potential matric see Matric potential ...

Potential osmotic see Osmotic potential ...

Potential pressure see Pressure potential ...

Potential water see Water potential ...

Potometry 7166, 7308, 7447, 7458

Precipitation, dew and canopy architecture 6661

Precipitation, dew and carbon dioxide influx 7246 7386

Precipitation, dew and chlorophyll 8117

Precipitation, dew and conductance for water vapour and carbon dioxide transfer 7228

Precipitation, dew and growth and productivity 6637, 6653, 6728, 6749, 6750, 6767,
      6804, 6845, 6883, 7010, 7081, 7085, 7086, 7088, 7090, 7303, 7318, 7499, 7500,
      7532, 7533, 7641, 7685, 7686, 7708, 7846, 7984, 7996, 8065, 8103

Precipitation, dew and leaf anatomy 7360, 7426

Precipitation, dew and respiration 7261

Precipitation, dew and stomata 7228, 7360, 7996

Precipitation, dew and transpiration 7261, 7334, 7360, 7996, 8012

Precipitation, dew and water absorption by plant 7246, 7998

Precipitation, dew and water status in plant 6707, 7360, 7907

Precipitation, dew and wilting 6952, 7614

Pressure bomb see Water potential, methods, pressure bomb

Pressure potential in plant tissue (see also Water status in plant ...) 6612,
      6660, 6681, 6726, 6746, 6770, 6782, 6784, 6864, 6879, 6900, 6901, 6910,
      6912, 6931, 7005, 7035, 7087, 7093, 7105, 7110, 7163, 7221, 7232, 7233, 7234,
      7259, 7281, 7312, 7350, 7351, 7352, 7419, 7444, 7469, 7476, 7494, 7495, 7576,
      7579, 7601, 7637, 7638, 7677, 7695, 7704, 7712, 7739, 7852, 7873, 7874, 7904,
      7905, 7906, 7922, 7940, 7990, 7995, 7996, 8013, 8073, 8080, 8088, 8123, 8124,
      8125, 8126, 8127,

Pressure potential, methods 7469, 7906, 8088, 8125, 8126

Productivity see Growth and productivity ...

Productivity of algae see Osmotically active substances ...; Salinity and produc-
      tivity of algae

Productivity of transpiration 6624, 6631, 6641, 6645, 6726, 6738, 6781, 6823, 6840,
      6875, 6907, 6998, 7027, 7104, 7128, 7162, 7170, 7176, 7177, 7179, 7192, 7195,
      7228, 7295, 7343, 7393, 7598, 7425, 7428, 7578, 7579, 7606, 7617, 7680, 7729,
      7771, 7805, 7820, 7876, 7877, 7878, 7906, 7918, 7943, 7992, 7996, 8010, 8011,
      8015

Proteins, amino acids, nucleic acids and water absorption by plant 6644, 6655, 6898,
      6928

Proteins, amino acids, nucleic acids and water status in plant   6741, 6746, 6762, 6787, 7048, 7118, 7233, 7390, 7394, 7395, 7786, 7900, 7907, 8049

Proteins, amino acids, nucleic acids and water transport in cells 7395

Proteins, amino acids, nucleic acids and wilting 6622, 6682, 6741, 6746, 6904, 6905, 6935, 7035, 7117, 7233, 7583, 7907, 7982, 7996, 8049, 8082

Psychrometry see Water potential, methods, psychrometry

R

Radiation see Irradiance ...

Rain see Precipitation, dew ...

Reactivity of stomata see Stomatal reaction rate; Stomatal reactivity during ontogenesis

Rehydratation 6611, 6622, 6648, 6649, 6688, 6689, 6714, 6786, 6853, 6896, 6900, 6970, 7006, 7011, 7035, 7193, 7234, 7288, 7351, 7432, 7449, 7554, 7599, 7610, 7719, 7723, 7835, 7985, 7996, 8035, 8070, 8073, 8113

Relative water content see Water saturation deficit

Resistance see Conductance ...

Respiration see Drought ...; Humidity of air ...; Irrigation ...; Osmotically active substances ...; Precipitation, dew ...; Salinity ...; Soil moisture ...; Water status in plant and respiration

Root pressure, exudation 6775, 6803, 6872, 6890, 6928, 7103, 7161, 7378, 7418, 7469, 7872, 8122

Root pressure, exudation, methods 8122

Root removal see Defoliation, decapitation, ear, root removal ...

Root, underground part and conductance for water vapour and carbon dioxide transfer 6911

Root, underground part and transpiration 6891

Root, underground part and water absorption by plant 6890, 6913, 7041, 7183, 7297, 7459, 7947, 7996

Root, underground part and water status in plant 6772, 6911, 6931, 7131, 7236, 7996,

Root, underground part and water transport in plant 7947

Root, underground part and wilting 6730, 7608

S

Saccharides see Carbohydrates ...

Saline water see Irrigation water quality

Salinity and canopy architecture 7026

Salinity and carbon dioxide influx 6618, 6693, 6752, 6885, 6888, 7113, 7175, 7237, 7266, 7292, 7409, 7482, 7541, 7576, 7628, 7694, 8023, 8048

Soil moisture and chloroplasts 7751

Soil moisture and conductance for water vapour and carbon dioxide transfer 6726,
     7002, 7104, 7130, 7170, 7442, 7449, 7544, 7621, 7626, 7739, 7996

Soil moisture and electron transport chain 7750, 7868, 8031

Soil moisture and growth and productivity 6608, 6614, 6625, 6631, 6642, 6653, 6659,
     6663, 6690, 6700, 6704, 6722, 6726, 6732, 6742, 6749, 6750, 6755, 6766, 6767,
     6770, 6777, 6807, 6816, 6824, 6829, 6834, 6847, 6851, 6860, 6862, 6883, 6945,
     6955, 6956, 6973, 6982, 6992, 7000, 7009, 7014, 7018, 7032, 7041, 7043, 7045.
     7056, 7064, 7068, 7069, 7083, 7084, 7091, 7098, 7102, 7109, 7112, 7117, 7134,
     7135, 7136, 7137, 7146, 7154, 7165, 7178, 7181, 7196, 7203, 7221, 7223, 7235,
     7236, 7238, 7239, 7251, 7264, 7265, 7280, 7312, 7313, 7318, 7334, 7345, 7348,
     7353, 7356, 7359, 7373, 7386, 7390, 7400, 7410, 7412, 7430, 7442, 7449, 7450,
     7452, 7457, 7471, 7472, 7480, 7493, 7496, 7498, 7503, 7508, 7532, 7534, 7539,
     7542, 7568, 7575, 7580, 7598, 7621, 7627, 7633, 7638, 7644, 7668, 7696, 7706,
     7708, 7709, 7727, 7730, 7765, 7766, 7799, 7801, 7804, 7820, 7833, 7846, 7849,
     7851, 7852, 7853, 7863, 7884, 7886, 7887, 7896, 7913, 7915, 7925, 7930, 7942,
     7949, 7951, 7959, 7960, 7962, 7969, 7970, 7971, 7975, 7985, 7994, 7996, 8013,
     8019, 8029, 8030, 8036, 8041, 8042, 8045, 8046, 8053, 8058, 8074, 8084, 8085,
     8103, 8110, 8111, 8118, 8119,

Soil moisture and leaf anatomy 6790, 6936, 7229, 7264, 7265, 7345, 7360, 7471, 7544,
     7566, 7598, 7600, 7733, 7985, 7996, 8101

Soil moisture and respiration 6726, 7134, 7386, 7421, 7655

Soil moisture and stomata 6912, 7259, 7360, 7449, 7864

Soil moisture and transpiration 6621, 6642, 6643, 6726, 6808, 6836, 6956, 7014, 7170,
     7230, 7247, 7360, 7377, 7386, 7399, 7410, 7434, 7435, 7457, 7471, 7598, 7600,
     7634, 7665, 7723, 7743, 7865, 7911, 7932, 7948, 7992, 7996, 8012, 8013

Soil moisture and water absorption by plant 7928

Soil moisture and water status in plant 6649, 6705, 6712, 6726, 6732, 6770, 6836, 6900
     6900, 7002, 7076, 7183, 7247, 7313, 7348, 7350, 7360, 7399, 7457, 7495, 7503,
     7579, 7588, 7626, 7665, 7739, 7857, 7905, 7990, 7996

Soil moisture and water transport in plant 6836, 7740

Soil moisture and wilting 6935, 6952, 7236, 7918, 7996

Soil moisture control, methods 6692, 6789

Soil moisture, methods 6692, 6735, 6814, 6837, 6902, 7122, 7202, 7211, 7372, 7589,
     7707, 7731, 7773, 7774, 7776, 7777, 7866, 7909, 7930, 7980, 7998, 8017, 8039

Soil water potential see Water potential in substrate

Solar radiation see Irradiance ...

Stomata see Altitude, Pressure ...; Carbohydrates ...; Carbon dioxide ...; Con-
     ductance for water vapour and carbon dioxide transfer ...; Cultivars ...;
     Drought ...; Ecotypes ...; Enzyme inhibitors ...; Enzymes ...; Farming
     practices ...; Flooding ...; Genetics ...; Growth substances, hormones,
     inhibitors etc. ...; Humidity of air ...; Irradiance ...; Irrigation ...;
     Leaf insertion level ...; Mineral elements ...; Mutants ...; Osmotically
     active substances ...; Oxygen ...; Pathogens ...; Pesticides, herbicides
     ...; pH ...; Pollutants, ozone ...; Precipitation, dew ...; Salinity ...;
     Soil moisture ...; Taxons ...; Temperature ...; Water status in plant ...;
     Wind and stomata

Structure of cuticle 6740, 6843, 6966, 6979, 7159, 7586, 7629, 7879, 8067, 8092

Structure of epidermis 6995, 7173, 7189, 7524, 7629, 7660

Sulphur oxides and other sulphur compounds see Pollutants, ozone ...

T

Taxons and conductance for water vapour and carbon dioxide transfer 6648, 6733, 6739, 6893, 6976, 7305, 7840, 7881, 7997, 8037

Taxons and stomata 6624, 6739, 6914, 6959, 7274, 7660

Taxons and transpiration 6643, 6648, 6739, 7088, 7283, 7511, 7593, 7945

Taxons and water absorption by plant 7378

Taxons and water status in plant 6648, 6739, 6893, 7076, 7184, 7712, 7793, 7840, 7945, 7997

Taxons and wilting 6648, 6660

Temperature and conductance for water vapour and carbon dioxide transfer 6620, 6641, 6647, 6658, 6717, 6726, 6842, 6934, 6998, 7008, 7092, 7104, 7170, 7209, 7269, 7341, 7351, 7371, 7419, 7490, 7501, 7502, 7526, 7544, 7556, 7572, 7578, 7645, 7739, 7742, 7809, 7881, 7939, 7955, 7996, 8037, 8127

Temperature and stomata 6717, 6792, 6998, 7092, 7093, 7209, 7360, 7545, 7645, 7721, 7809, 7939, 7946, 7996, 8127

Temperature and transpiration 6641, 6712, 6726, 6825, 6887, 6956, 6998, 7008, 7021, 7105, 7158, 7170, 7192, 7198, 7349, 7351, 7360, 7371, 7389, 7480, 7526, 7544, 7545, 7567, 7578, 7620, 7649, 7696, 7710, 7877, 7918, 7932, 7996, 8051, 8056

Temperature and water absorption by plant 6706, 6872, 6887, 7349, 7594, 7691, 7874

Temperature and water status in plant 6620, 6666, 6726, 6736, 6842, 6923, 7048, 7092, 7093, 7158, 7336, 7360, 7371, 7419, 7742, 7922, 8018, 8090

Temperature and water transport in cells 6831, 7255, 7363, 7591, 7673, 7769, 7996

Temperature and water transport in plant 7742, 8002

Temperature and wilting 6640, 6736, 6887, 7349, 7594, 7599, 7710, 7918, 7996, 8018

Transpiration see Age of plant ...; Anatomical structure ...; Carbohydrates ...; Cultivars ...; Defoliation, decapitation, ear, root removal ...; Deuterium oxide, tritium oxide ...; Drought ...; Ecotypes ...; Farming practices ...; Flooding ...; Genetics ...; Growth substances, hormones, inhibitors etc. ...; Humidity of air ...; Irradiance ...; Irrigation ...; Leaf insertion level ...; Mineral elements ...; Osmotically active substances ...; Oxygen ...; Pathogens ...; Pesticides, herbicides, ...; Pollutants, ozone ...; Precipitation, dew ...; Salinity ...; Soil moisture ...; Taxons ...; Temperature ...; Water status in plant ...; Wind and transpiration

Transpiration chambers 6623, 6734, 7275, 7556

Transpiration coefficient 6612, 6619, 6629, 6633, 6642, 6643, 6647, 6748, 6805, 6839, 6853, 6883, 7014, 7043, 7070, 7074, 7084, 7099, 7111, 7135, 7139, 7140, 7146, 7227, 7228, 7312, 7334, 7451, 7455, 7479, 7544, 7545, 7603, 7612, 7643, 7680, 7752, 7801, 7804, 7828, 7845, 7884, 7897, 7918, 7991, 7992, 7993, 7996, 8036, 8065, 8080, 8081, 8118

Transpiration curves 6660, 6891, 6924, 7399, 7574, 7609, 7618, 7657, 8040, 8061

Transpiration cuticular 6793, 7599, 7525, 7557, 7657, 8040

Transpiration integrated 6932, 6956, 7049, 7138, /228, 7620, 7936, 8013, 8014, 8043,
        8051

Transpiration rate and antitranspirants see Antitranspirants

Transpiration rate and fine structure 7571

Transpiration rate and stomata see Stomata and transpiration rate

Transpiration rate and temperature 7343

Transpiration rate, comparison of plants with different types of carbon metabolism
        7389

Transpiration rate, diurnal changes 6647, 6658, 6712, 6726, 6738, 7002, 7047, 7088,
        7089, 7224, 7259, 7302, 7351, 7352, 7424, 7425, 7457, 7474, 7526, 7545, 7565,
        7620, 7634, 7649, 7677, 7683, 7/11, 7723, 7771, 7772, 7911, 7932, 7936, 7978,
        7979, 8028, 8051, 8079, 8080

Transpiration rate, gradients of air humidity in leaf intercellulat spaces 6649,
        6739, 7253, 8000

Transpiration rate, heterogeneity of single leaf blade 7282, 8010

Transpiration rate in artificial conditions 6618, 6623, 6624, 6648, 6658, 6662, 6685,
        6698, 6737, 6739, 6771, 6808, 6823, 6825, 6876, 6886, 6887, 6891, 6907, 6924,
        6934, 6956, 7008, 7021, 7027, 7105, 7163, 7166, 7170, 7192, 7253, 7268, 7270,
        7295, 7300, 7349, 7365, 7382, 7386, 7431, 7447, 7463, 7480, 7526, 7553, 7571,
        7572, 7579, 7596, 7600, 7603, 7604, 7607, 7609, 7612, 7617, 7634, 7636, /653,
        7665, 7683, 7689, 7723, 7744, 7747, 7870, 7934, 7936, 7954, 7961, 7972, 7978,
        7992, 7997, 8002, 8011, 8015, 8021, 8028, 8051, 8076, 8079, 8080

Transpiration rate in natural conditions 6646, 6647, 6684, 6712, 6726, 6738, 6755,
        6811, 6836, 6837, 6915, 6920, 6993, 7002, 7092, 7158, 7187, 7259, 7310, 7311,
        7350, 7352, 7356, 7377, 7379, 7397, 7399, 7425, 7455, 7568, 7620, 7638, 7640,
        7643, 7656, 7677, 7741, 7749, 7762, 7876, 7911, 7924, 7930, 7935, 7936, 7945,
        7954, 7979, 7992, 8001, 8015, 8043

Transpiration rate, methods, gasometric systems 6623, 6685, 6729, 6818, 6886, 7275,
        7341, 7556, 7807, 8051, 8076

Transpiration rate, methods, gravimetric 7744, 7747

Transpiration rate, methods, infra-red gas analysers

Transpiration rate, methods, other hygrometers 6734

Transpiration rate, oscillations   7170, 7225, 7600, 7860

Transpiration rate, seasonal changes   6647, 6711, 6726, 6853, 7002, 7080, 7158,
        7162, 7261, 7302, 7311, 7316, 7318, 7379, 7416, 7424, 7435, 7464, 7486, 7536,
        7543, 7593, 7649, 7657, 7677, 7711, 7720, 7759, 7934, /943, 7979, 7996

Transpiration rate, theoretical background 6726, 7114, 7115, 7474, 7497, 7515, 7649,
        7650, 7681, 7720, 7747, 7749, 7764, 7791, 7801, 7880, 7894, 7932, 7979

Transpiration stomatal 7095, 7525, 7557, 7657, 8040

Transport of water see Water transport ...

Trichomes   see   Leaf surface, waxes and trichomes

Turgor pressure   see   Pressure potential ...

W

Water absorption and ion uptake 6674, 6766, 6863, 6951, 6003, 7197, 7217, 7220, 7280, 7287, 7442, 7611, 8063, 8069

Water absorption by discs of leaf tissue 8004

Water absorption by parts of plant 6657, 6924, 7740, 7767, 7768, 7966, 8021

Water absorption by plant see Age of plant ...; Cultivars ...; Deuterium oxide, tritium oxide ...; Drought ...; Enzyme inhibitors ...; Farming practices ...; Flooding ...; Genetics ...; Growth substances, hormones, inhibitors etc. ...; Humidity of air ...; Irradiance ...; Irrigation ...; Mineral elements ...; Osmotically active substances ...; Oxygen ...; Pathogens ...; Pesticides, herbicides ...; Precipitation, dew ...; Proteins, amino acids, nucleic acids ...; Salinity ...; Soil moisture ...; Taxons ...; Temperature and water absorption

Water absorption by plant, diurnal changes 7376

Water absorption by plant, seasonal changes 6661, 6700, 6701, 6711, 6773, 7126, 7138, 7423, 7971, 8093

Water absorption by seeds 6622, 6644, 6655, 6677, 6688, 6689, 8778, 6785, 6813, 6816, 6861, 6880, 6898, 6906, 6971, 7142, 7143, 7147, 7152, 7200, 7294, 7337, 7340, 7373, 7439, 7481, 7624, 7725, 7727, 7785, 7788, 7824, 7826, 7838, 7847, 7850, 7874, 7916, 7923, 8006, 8008, 8070, 8095

Water absorption from atmosphere 6706, 7020, 7042, 7246, 7487, 7641, 7656, 7679, 7748

Water absorption from soil 6610, 6695, 6766, 6836, 6913, 7020, 7021, 7041, 7066, 7162, 7166, 7183, 7236, 7242, 7247, 7275, 7340, 7376, 7424, 7433, 7457, 7475, 7579, 7641, 7845, 7881, 7947, 7948, 7964, 7996, 8044

Water absorption from solution 6798, 6887, 7349, 7458, 7554, 7653, 7664, 7691, 8063

Water balance of cells and tissues 7621

Water balance of whole plant 6887, 7110, 7349, 7614, 7653, 7747, 8043

Water balance of whole plant, methods 7349, 8112

Water consumption 6627, 6631, 6711, 6738, 6739, 6867, 6963, 6969, 7004, 7139, 7140, 7145, 7230, 7317, 7368, 7644, 7706, 7792, 7805, 7832, 7853, 7855, 7897, 7901, 7915, 7988, 7992, 7993, 8081, 8118, 8119

Water consumption, methods 6616, 6963, 7138, 7277, 7317, 7537, 7828, 7886, 7901, 7915, 8112

Water content in plant and related volume changes 7004, 7108, 7350, 7351, 7446, 7621, 7696, 7723, 7745, 7785, 7848, 7940, 8064, 8124

Water content in plant, methods 6753, 6754

Water status in plant, diurnal changes 6609, 6632, 6635, 6681, 6712, 6726, 6769,
        6836, 6910, 7002, 7047, 7092, 7093, 7120, 7184, 7185, 7241, 7247, 7248, 7259,
        7293, 7352, 7377, 7501, 7502, 7576, 7581, 7582, 7588, 7612, 7621, 7677, 7695,
        7723, 7735, 7739, 7740, 7747, 7840, 7857, 7905, 7918, 7935, 7937, 7939, 7949,
        7954, 7990, 7996, 7997, 7999, 8093, 8098

Water status in plant, heterogeneity of single leaf blade 7734

Water status in plant, oscillations 7554

Water status in plant, seasonal changes 6625, 6626, 6697, 6705, 6726, 6767, 6893,
        6911, 6926, 7093, 7156, 7222, 7241, 7247, 7267, 7301, 7320, 7325, 7415, 7446,
        7503, 7612, 7621, 7647, 7677, 7695, 7712, 7739, 7741, 7743, 7775, 7793, 7794,
        7840, 7895, 7932, 7935, 7945, 7990, 8005, 8054, 8093, 8106

Water status in plant, theoretical background, terminology 7312, 7364, 7996

Water stress develoment see Drought and wilting

Water stress in plant see Wilting ...

Water transport in cells (see also Carbohydrates ...; Carbon dioxide ...; Deu-
        terium oxide, tritium oxide ...; Drought ...; Growth substances, hormones,
        inhibitors etc. ...; Lipids, fatty acids ...; Mineral elements ...; Osmo-
        tically active substances ...; Pollutants, ozone ...; Proteins, amino acids,
        nucleic acids ...; Salinity ...; Temperature ...; Water status in plant and
        water transport in cells) 6708, 6709, 7046, 7132, 7249, 7255, 7284, 7292,
        7314, 7592, 7791, 7831, 8126

Water transport in cells, cell wall structure, modulus of elasticity 6864, 6896,
        6899, 6910, 7234, 7245, 7495, 7506, 7673, 7769, 7864, 7906, 7940, 7996, 8016,
        8126

Water transport in cells, membrane structure 7323, 7418, 7531, 7591, 7630, 7673,
        7714, 7769, 7950, 7996, 8123

Water transport in cells, methods 7592

Water transport in cells, permeability 6831, 6864, 7067, 7174, 7255, 7363, 7439,
        7445, 7592, 7819, 7864, 7906, 7937

Water transport in cells, plasmolysis 6672, 7057, 7067, 7093, 7245, 7257, 7358, 7591,
        7698, 7699, 7786, 7950, 8124

Water transport in cells, vacuole development 8063, 8110

Water transport in plant see also Age of plant ...; Altitude, pressure ...; Cul-
        tivars ...; Deuterium oxide, tritium oxide ...; Growth substances, hormones,
        inhibitors etc. ...; Humidity of air ...; Irradiance ...; Mineral elements
        ...; Pesticides, herbicides ...; Salinity ...; Soil moisture ...; Tem-
        perature ...; Water status in plant and water transport in plant

Water transport in plant, conductances 6836, 6887, 6891, 7447, 7458, 7734, 7740,
        7742, 8126

Water transport in plant, diurnal changes 6836, 7475, 7749, 7936, 8098

Water transport in plant, methods 6813

Water transport in plant, radial transport in tree stems 7871

Water transport in plant, transport in leaf 6908, 7458, 7734, 8000, 8098, 8125

Water transport in plant, transport in other organs than above and below 7208, 7734, 8098

Water transport in plant, transport in root 6872, 6891, 7236, 7309, 7310, 7417, 7447, 7458, 7459, 7600, 7734, 7740, 7837, 7947

Water transport in plant, transport in xylem, methods 7747, 7749, 8043, 8098

Water transport in plant, transport in xylem of herbaceous stem 6957, 7310, 7604, 7749, 8002

Water transport in plant, transport in xylem of tree stem 6836, 6837, 7475, 7740, 8098, 8107

Water transport in plant, transport soil - root 6695, 6819, 7247, 7281, 7310, 7418, 7600, 7739, 7740, 7947, 8044

Water transport in plant, vascular bundle structure 6681, 6991, 7123, 7309, 7639, 7749

Water transport in soil 6610, 6695, 6701, 6819, 6865, 6982, 7211, 7217, 7236, 7243, 7281, 7424, 7539, 7600, 7658, 7664, 7740, 7815, 7981, 7981, 7996, 8017, 8044

Waxes see Leaf surface, waxes and trichomes; Leaf surface, waxes and trichomes, seasonal changes

Wilting see Age of plant ...; Anatomical structure ...; Antibiotics ...; Carbohydrates ...; Cultivars ...; Defoliation, decapitation, ear, root removal ...; Deuterium oxide, tritium oxide ...; Drought ...; Ecotypes ...; Enzymes ...; Farming practices ...; Genetics ...; Growth substances, hormones, inhibitors etc. ...; Humidity of air ...; Irradiance ...; Irrigation ...; Leaf insertion level ...; Mineral elements ...; Osmotically active substances ...; Oxygen ...; Pathogens ...; Pesticides, herbicides ...; Pollutants, ozone ...; Precipitation, dew ...; Proteins, amino acids, nucleic acids ...; Salinity ...; Soil moisture ...; Taxons ...; Temperature and wilting

Wilting and anatomical structure 7453, 7996, 8062

Wilting and ion absorption and transport 6617, 7016, 7233, 7665, 7925, 7977, 8068

Wilting and other processes than above and below 6607, 661, 6634, 6654, 6667, 6672, 6678, 6710, 6721, 6723, 6724, 6758, 6759, 6761, 6782, 6850, 6856, 6857, 6879, 6897, 6929, 6941, 6980, 6983, 7013, 7024, 7035, 7050, 7052, 7054, 7108, 7116, 7175, 7233, 7273, 7290, 7299, 7384, 7414, 7415, 7439, 7440, 7445, 7477, 7483, 7564, 7569, 7570, 7582, 7625, 7626, 7661, 7671, 7676, 7722, 7756, 7798, 7825, 7874, 7904, 7977, 8003, 8030, 8064, 8068, 8089

Wilting and plant metabolism 6640, 6648, 6678, 6721, 6786, 6820, 6821, 6878, 7116, 7119, 7175, 7233, 7458, 7569, 7904, 7965, 7982, 8003

Wilting, diurnal changes 7212, 7352, 7614, 7918

Wilting, indicators, methods 6616, 7141, 7157, 7496, 7886, 7901

Wilting, mechanismus of development, indicators 6611, 6634, 6635, 6648, 6649, 6656, 6665, 6681, 6714, 6721, 6730, 6769, 6808, 6822, 6847, 6900, 6901, 6905, 6912, 6952, 7006, 7011, 7013, 7105, 7108, 7116, 7118, 7193, 7231, 7232, 7233, 7234, 7238, 7358, 7384, 7432, 7457, 7483, 7509, 7578, 7582, 7583, 7610, 7661, 7712, 7723, 7798, 7822, 7835, 7858, 7859, 7898, 7904, 7932, 7973, 7996, 8003, 8061, 8072, 8113

Wilting, oscillations 6665

Wind and conductance for water vapour and carbon dioxide transfer 7078, 7829, 7876,
     7918

Wind and stomata 7918

Wind and transpiration 6726, 7198, 7392, 7877, 7918, 7932

Wind and water status in plant 7918

X

Xylem transport of water see Water transport in plant, transport in xylem ...